# Intelligent Interpretation for Geological Disasters

Weitao Chen · Cheng Zhong · Xuwen Qin ·
Lizhe Wang

# Intelligent Interpretation for Geological Disasters

## From Space-Air-Ground Integration Perspective

Springer

Weitao Chen
School of Computer Science
China University of Geosciences
Wuhan, China

Xuwen Qin
Natural Resources Aerospace Remote
Sensing Centre
China Geological Survey
Beijing, China

Cheng Zhong
Badong National Observation
of Geohazards
China University of Geosciences
Wuhan, China

Lizhe Wang
School of Computer Science
China University of Geosciences
Wuhan, China

ISBN 978-981-99-5821-4 ISBN 978-981-99-5822-1 (eBook)
https://doi.org/10.1007/978-981-99-5822-1



This Springer imprint is published by the registered company Springer Nature Singapore Pte Ltd.
The registered company address is: 152 Beach Road, #21-01/04 Gateway East, Singapore 189721, Singapore

# Foreword

Currently, climate change, human activities, and tectonic movement contribute to the increasing frequency and severity of geological disasters around the world. These events can have severe and far-reaching consequences including loss of human life, damage to infrastructure and property, and environmental degradation. However, the intermittency and unpredictability of geological disasters make early warning and preparation very challenging. In addition, the severity of geological disasters can be influenced by a range of factors, including the location and intensity of geological events, the population density and infrastructure of affected areas, and the effectiveness of integrative emergency responses. Therefore, it is critical to strengthen the monitoring and mitigation of risks associated with geological disasters to minimize their impacts on human life and surrounding environment.

In modern earth science research, the application of remote sensing (RS) technology is essential for effectively monitoring and management of geological disasters. Traditional RS methods to study geological disaster rely on detecting variations in land cover and topography, as well as ground displacement and deformation. These methods involve visual interpretation and manual analysis, leading to time-consuming processes that are subject to human error and cannot provide results of near real-time monitoring. The development of earth science increasingly relies on technological innovation in recent decades, and the scientific research paradigm is undergoing significant changes with interdisciplinary integration, which is particularly the case for the fields of observation, detection, and simulation technologies for high-precision, multi-scale big data. The integration of the new generation artificial intelligence and earth science can develop new research methodologies and technical tools for geological disasters studies. The intelligent interpretation of RS data of geological disasters involves four primary disciplines: earth science, remote sensing science, computer science, and intelligence science.

The lead author of this book, Prof. Dr. Weitao Chen, my former Ph.D. student, has been actively involved in RS studies focusing on geo-environment, and this book was designed to summarize the results of his group's recent work on geological disaster and deep learning. It covers a range of topics from geological disaster detection and monitoring to the susceptibility assessment, with slope geological hazard and ground fissures as the main focuses. Additionally, this book offers a detailed presentation of the principles and methods of utilized intelligent technology, constructs datasets for multi-scale and multi-type geological disasters, and forms a theoretical framework system for intelligent interpretation of RS data for geological disaster study.

This book can guide the new generation of geoscientists to utilize artificial intelligence to expedite the transformation of deep learning theoretical research into practical applications in geological disaster prevention and control. I hope that this book will serve as a reference for researchers, practitioners, and policymakers who are interested in the application of artificial intelligence and remote sensing in geological disaster monitoring and management.

Yanxin Wang
Professor
China University of Geosciences
Wuhan, China

Member
Chinese Academy of Sciences
Wuhan, China

# Preface

Geological disasters can generate significant socio-economic and environmental influences, which lead to loss of life, destruction of infrastructure, disruption of essential services, and environmental degradation. Although ground-based and remote sensing techniques have been widely used in the past a few decades, the challenges of time-consuming processes, subjection of human error, have not been well addressed yet, which thereby demand the development of intelligent interpretation methods to improve efficiency and performance. This book constructs multi-type geological disaster datasets and carries out the researches on the theories and methods for the intelligent interpretation of geological disasters based on ground-based, airspace-based, and space-based data.

Focusing on slope hazards and ground fissures in China, the book studies the five aspects: (1) the intelligent analyzing methods for the prediction of landslide displacement based on long-term ground monitoring data; (2) the deep learning-based methods to identify landslide from optical satellite remote sensing images; (3) the deep learning-based methods for the satellite monitoring of landslide evolution and changes based on time series optical satellite remote sensing images; (4) the intelligent assessment methods for landslide susceptibility; and (5) the deep learning-based methods to recognize ground fissures based on aerial remote sensing images.

Chapter 1 was written by Weitao Chen and Xuwen Qin and assisted by Hangyuan Liu and Shubing Ouyang. Chapter 2 was written by Xuwen Qin and Ruizhen Wang. Chapter 3 was written by Ruizhen Wang and assisted by Qinyao Zhu. Chapter 4 was written by Cheng Zhong and assisted by Ruizhen Wang. Chapter 5 was written by Weitao Chen and Xuwen Qin, with the assistance of Yuqing Xiong and Ruizhen Wang. Chapter 6 was written by Xuwen Qin and assisted by Wenyue Sun and Weitao Chen. Chapter 7 was written by Zixin Liu, Weitao Chen, Xubo Gao and Lizhe Wang. The experiments of the book were designed by Weitao Chen, Cheng Zhong, Xuwen Qin, and Lizhe Wang and completed by Qinyao Zhu, Zixin Liu, Wenyue Sun, and Yuqing Xiong. The work of the whole book was completed by Weitao Chen, Cheng Zhong, Xuwen Qin, and Lizhe Wang.

This book was jointly supported by the key research and development program of Hubei province (No. 2021BID009) and the Fundamental Research Funds for the Natural Science Foundation of China (No. U21A2013).

The book is intended for senior undergraduate students, postgraduate students, and Ph.D. students who are interested in geological disasters, natural hazards, remote sensing, and artificial intelligence. It can also be used as a reference book for researchers, practitioners, and policymakers to guide the geological disaster monitoring and management.

Wuhan, China

Weitao Chen
Cheng Zhong
Xuwen Qin
Lizhe Wang

# Contents

# Chapter 1
# Geological Disaster: An Overview

**Abstract** Geological disasters are catastrophic events that result from natural processes in the Earth's crust, including earthquakes, landslides, rock falls, ground fissures, etc. These disasters can cause significant damage to infrastructure, property, and human lives, as well as impact the environment and natural resources. This chapter provides an overview of the origin and types of main terrestrial geological disasters, and introduces the risk assessment and management strategies used for geological disaster researches by using equipment-based and remote sensing-based approaches. Key findings and insights from recent researches, as well as gaps and challenges in current research and management efforts are also discussed. The futural directions and opportunities presented in this chapter can help to inform stakeholders and researchers on the potential approaches for managing and mitigating the impact of geological disasters.

## 1.1 Description of Geological Disaster

Geological disasters refer to natural disasters that occur in the geological environment and are related to the properties, structures and functions of geological bodies, which can lead to serious influences on human activities. Many factors such as tectonic movement, climate change, and human activities can lead to geological disasters, and their categories mainly include earthquakes, debris flows, landslides, collapses, land subsidence, and ground fissures (Kusky, 2003). This section will discuss in detail the causes and types of geological disasters.

### *1.1.1 Origin of Geological Disaster*

The causes of geological disasters are complex and can be categorized to factors inside and outside the earth. The following parts will discuss the causes of geological disasters from the two aspects separately.

**The Earth's Interior Factors**



W. Chen et al., *Intelligent Interpretation for Geological Disasters*,
https://doi.org/10.1007/978-981-99-5822-1_1

(1) Plate tectonics: The friction and collision generated between the plates will cause geological disasters such as earthquakes and volcanic eruptions. For example, the Pacific Rim of Fire is one of the most active volcanic belts on the earth, and its formation and activity are closely related to the interaction of the Pacific plate and other plates (Bird & Liu, 2007; Kusky, 2008).
(2) Geological structure: Geological structure is a structural form inside the earth, including mountains, uplifts, faults, ground fissures, etc. The deformation and movement of these geological structures can also cause disasters such as earthquakes, landslides, and collapses. For example, China's seismic zones are mainly distributed at the junction of the Eurasian continental plate and the Pacific plate, where the geological structure is relatively complex and seismic activity is frequent (Chen et al., 2009).
(3) Volcanic activity: Volcano activity is a phenomenon within the earth where volcanic eruptions and lava inflows lead to the deformation and destruction of lands. The accumulation of volcanic ash and rock can also generate geological disasters in surrounding areas. For example, the eruption of Mount Merapi in Indonesia caused massive landslides and debris flows (Sun et al., 2022).
(4) Geothermal activity: Geothermal activity is a form of energy release from the interior of the earth, including volcanic heat, hydrothermal fluid, and geothermal heat. Surface deformation and destruction caused by geothermal activities can also lead to geological disasters. For example, there are large-scale geothermal activities in the Qinghai Lake area in China, which has caused disasters such as ground fissures and earthquakes (Chen et al., 2022).

**The Earth's Surface Factors**

(1) Stability of rock and soil mass: Rock and soil mass are the geomaterials on the surface of the earth. When they are subjected to external forces or they are of low inherent shear strength in some regions, it may form landslides and collapses here (Zhang et al., 2021). For example, the Loess Plateau in China often suffers from large-scale landslides and mudslides due to the fragile soil.
(2) Hydrogeological factors: Hydrogeological factors refer to the impact of surface water and groundwater on the occurrence of geological disasters (De Luca & Versace, 2017). For example, extreme precipitation will increase the rainfall runoff and the dynamics of surface water, thereby may cause debris flows in mountainous areas.
(3) Human activities: Human engineering activities such as mining, highway and reservoir construction will affect the geological environment on the surface of the earth and cause geological disasters (Li et al., 2020). In addition, the process of urbanization enhances the over-exploitation of groundwater resources, leading to the ground subsidence (Chen et al., 2012).
(4) Climate factors: Climate warming will lead to glacier retreat and mountain collapse (Dhakal, 2015). And drought extremes will lead to water and land loss, and desertification. For example, global warming has led to the rapid melting of the ice on Greenland, accelerating the sinking of the island and rising of sea levels, further triggering earthquakes and tsunamis (Dong et al., 2022).

(5) Chain effects of geological disasters: There are often interrelated and chain effects between natural disasters, such as earthquakes can cause landslides and collapses, while the lasting precipitation extremes can enhance the debris flows (Xu et al., 2014).

### *1.1.2 Types of Geological Disaster*

Geological hazards can be classified into two categories of mass movements and ground instability. Mass movements include landslide, debris flow and rockfall. Ground instability includes ground subsidence, sinkhole, and collapse.

**Mass Movements**

(1) Landslide

Landslide refers to the phenomenon that the surface or underground rock mass slides or flows downward along a certain surface or surface group under the external forces. Landslides are usually caused by geological structure deformation, groundwater gushing, rainfall and other factors. Based on the geomaterials, the types of landslides include rockfall landslides, rock landslides, mudslides and snow avalanches, etc. Rockfall landslides are caused by the loose or fractured rocks on the hillside. Long-term weathering and erosion allow the rock to gradually loosen and lose the carrying capacity, which will then collapse and slide when subjected to external forces. Rock landslides are usually caused by the different lithologies in the mountain, such as layered rocks with alternating hardness and softness, rocks with developed joints, etc. Mudslides are easily caused when the soil contains higher proportion of clay. During the rainy time, the clay soil will become fluid and form landslides. Snow avalanches are caused by the melting of ice and snow on high mountains to form flowing water with covered geomaterials. When the slope angle is large enough, the speed of the flow increases significantly and forms the avalanche (Causes, 2001).

(2) Debris flow

Debris flow refers to a sudden mountain torrent disaster caused by the destruction and disturbance of the surface ground materials in mountainous areas due to flash floods, precipitation extremes or snowmelt. Debris flow has the characteristics of fast speed, large volume, strong erosion and destructive power. The process of debris flow formation can be separate into preparatory stage, developing stage and fading stage. In the preparatory stage, a large amount of debris flow materials are accumulated in the mountainous area, but the debris flow has not yet formed. During the developing stage, debris flow moves downstream due to certain trigger conditions. The fading stage refers to the gradual fade away of the debris flow and stops its development.

The formation of debris flow requires certain geological conditions, such as steep hillsides, rocks that are prone to collapse, and large amounts of rainfall. In addition, human activities may also have a certain impact on the formation of debris flows, such as over-harvesting and unreasonable construction, etc. The impact range of debris flows can be very wide, and may involve roads, bridges, buildings in mountainous areas, causing serious damage to surrounding residents and farmland. In addition, it may form deposit when debris flows move through low-lying areas, leading to additional floods and other disasters (Takahashi, 1981).

(3) Rockfall

When rocks on mountains or cliffs loosen or break, they may roll or fall from the heights, and cause harm to surrounding creatures. Rockfall is generally related to the geological structure of the mountain, rock properties, topography, climate and hydrology. The suddenness, unpredictability, and instantaneity of rockfall disaster which pose a threat to infrastructure such as transportation, hydropower, and personnel and property safety in mountainous areas.

Rockfall disasters can be classified into single rockfall, rock landslide, rock avalanche. Single rockfall refers to a single rock or gravel rolling down from the mountain, which may lead to traffic accident and casualties on mountain roads and railway lines. Rock landslide refers to the sliding of a large amount of rock mass along the slope surface. This scale of rockfall usually occurs on a steep hillside, and it is easily to cause deposits, displacement of the river, and damage to the slope surface. Rock avalanches refer to the downward sliding or collapse of rocks due to the rupture of rock formations or structural planes (Vandewater et al., 2005).

**Ground Instability**

(1) Ground subsidence

Ground subsidence hazards refer to the decline of the ground surface elevation relative to the reference elevation, usually caused by over-exploitation of groundwater, mining of underground mineral resources, seismic activity and natural deposition. The groundwater level decrease causes the underground soil layer to shrink and loss water content, resulting in a reduction in the volume of the soil layer, thereby leading to the ground subsidence. And during the mining process, soil layers are moved or weakened, also causing the soil layer to shrink. Earthquakes cause deformation and destruction of soil layers, causing the ground to drop. Natural deposition and human activities may also lead to ground subsidence, such as river sedimentation and construction of urban groundwater and subway systems (Bagheri-Gavkosh et al., 2021).

Ground subsidence will deform the structure of the building and generate cracks in the piping system. And the deformation of surface substance will bring many negative effects to agriculture, ecological environment and urban planning.

(2) Sinkhole

Sinkholes are formed as a result of underground caves or karst landforms, which are formed by chemical reactions in groundwater. Over time, these chemical reactions weaken the rock and soil beneath, eventually causing the ground to collapse and form sinkholes. Sinkhole hazards typically occur in areas with high water tables, such as near rivers, lakes and coastlines. In these areas, the water table can erode rock and soil, forming the underground cavities. When the pressure on the ground exceeds the support capacity of the surface materials, the ground will collapse to form a sinkhole (Caramanna et al., 2008).

Sinkhole disasters can lead to casualties, house collapse, road damage, underground pipeline rupture, water pollution, but they are usually difficult to predict and control.

(3) Ground fissure

Ground fissure is a type of geological feature that occurs when the ground surface separates or cracks apart due to movement or displacement of the surface substance, like rock or soil. These movements can be caused by tectonic activity, landslides, ground subsidence, or the settlement of underlying bedrock. Ground fissures can range in size from just a few centimeters to several meters in width and length, and can be several meters deep. They typically follow a linear or curvilinear pattern, and can extend for a number of kilometers (Da-yu & Li, 2000).

Ground fissures can have a variety of effects on the surrounding area. In some cases, they may simply be a visual disturbance. In other cases, ground fissures can lead to significant damage to surrounding constructions, roads, bridges, especially if the fissure widens or deepens over time.

## 1.2 Risk Assessment and Management

### *1.2.1 Disaster Assessment*

Geological disasters have brought a huge threat to the safety of human life and property, so the risk assessment and management of geological disasters can be very important. Risk assessment refers to the assessment of the possible impact and probability of a disaster. In geological disaster risk assessment, it is firstly necessary to collect relevant geological, topographical, hydrological data, and then conduct analysis and comprehensive quantified assessment to determine the probability of disaster occurrence and possible impact on the surrounding environment and human beings, so that effective prevention and control management strategies can be formulated for disaster mitigation.

So far, there have been some popular geological hazard assessment methods and tools developed to predict and assess the occurrence and impact of geological hazards, and to take corresponding measures for risk management and disaster reduction (Fanos & Pradhan, 2018; Hürlimann et al., 2008; Pardeshi et al., 2013; Theron & Engelbrecht, 2018).

**Geological Disaster Assessment Methods**

(1) Quantitative assessment methods: Including statistical methods, modelling methods, probability methods, etc. These methods are mainly based on historical data and statistical analysis to assess the risk of geological disasters. The

modelling method is based on field investigation and experimental data, and predicts the probability of geological disasters by constructing physical or mathematical models (Wang et al., 2021b).

(2) Qualitative assessment method: Using expert experience, investigation, and other research methods to conduct qualitative analysis on the risk of geological disasters. It mainly includes judgment research and engineering empirical method (Wang et al., 2013).

**Geological Disaster Assessment Tools**

(1) GIS tools: Geographic Information System (GIS) is a computer system that integrates data and analysis functions and can be used for geological disaster assessment. It can help analyze factors such as terrain, soil, geological structure, and the occurrence of historical geological disasters. Disasters can be assessed by generating geological disaster risk and vulnerability maps (Wang et al., 2021a).
(2) Remote sensing technology: Remote sensing technology can obtain a large amount of surface information in the forms of satellite images, lidar data, etc. These data can be used to study the formation mechanism of geological disasters and the influence area of disaster. The development trend of geological hazards can also be analyzed by comparing multi-temporal satellite images (Jiang et al., 2017).
(3) Numerical simulation tools: Numerical simulation tools can simulate the process of geological disasters by constructing physical models of geological disasters. These models can be used to predict the possible occurrence time and scale of geological disasters, and to evaluate the effectiveness of different disaster prevention measures (Li et al., 2021).
(4) Intelligent decision-making system: Intelligent decision-making system is a software tool based on artificial intelligence technology, which can be used for prediction and early warning of geological disasters. It can automatically evaluate the geological disaster risk by the analysis and processing of multi-source data, and provide corresponding early warning information and strategies (Liu et al., 2015).

**Limitations of Current Geological Disaster Assessment**

Currently, there are still some limitations and challenges in the current geological hazard risk assessment though with abovementioned methods and tools, including the following aspects:

(1) Insufficient data: Geological disaster risk assessment requires a large amount of geological, topographic, meteorological and other data, but the acquisition of these data is costly, especially for some relatively unachievable regions, which leads to a lower accuracy of assessment results (Sun et al., 2020).
(2) Uncertainty: The results of geological disaster risk assessment are usually affected by many uncertain factors, such as earthquakes, rainfall, etc. All the

assessments are the simplified natural processes that cannot include all factors, which will lead to errors (Beven et al., 2018).

(3) Lack of unified standards: Currently, the standards and methods of geological disaster risk assessment are different, and the lack of unified standards makes the comparability of assessment methods insignificant (Wu et al., 2021a).
(4) Geological hazards are frequent: Different types of geological disasters can interact with each other and generate chain effects, so multiple factors need to be considered in the assessment simultaneously, such as earthquakes, landslides, debris flows, etc., which increases the complexity of the assessment (Zhou et al., 2015).
(5) Human factors: Human activities will also affect the occurrence and risk of geological disasters, such as mining and construction, but these factors are difficult to be quantified and will increase the difficulty of disaster assessment (Li et al., 2020).

### *1.2.2 Risk Management*

#### Management Methods of Geological Disasters

After effective assessment, managing and mitigating geological disasters will be the following task. There are some strategies and methods that are widely used to manage and mitigate geological disasters in recent days (Chae et al., 2017; Nathe, 2000):

(1) Monitoring and early warning: By monitoring changes in geological disaster areas in real time, early warnings can be made to avoid or mitigate losses caused by geological disasters.
(2) Engineering management: The occurrence and development of geological disasters can be effectively controlled by using different engineering methods, such as earthwork reinforcement, rock mass reinforcement, slope protection, embankment protection, etc.
(3) Land use control: Reasonable land use control can reduce ground damage and reduce the probability of geological disasters.
(4) Public education: The occurrence of geological disasters is often due to the lack of public awareness and prevention awareness of geological disasters. The occurrence and loss of geological disasters can be reduced by carrying out publicity and education activities to improve public awareness of geological disasters.

#### Limitations of Current Geological Disaster Management

However, the effective management of geological disasters still faces many challenges and obstacles (Menon, 2012), mainly including the following problems:

(1) Data acquisition and analysis: Effective disaster management requires a large amount of geological, topographic, meteorological, hydrological data, and the

acquisition, integration and analysis of these data is a complex process. In addition, the quality of these data may also vary, which can affect the accuracy of risk assessment and decision-making.

(2) Uncertainty in risk management: In the process of disaster management, uncertainty comes from many aspects, including data quality, model accuracy, unknown variables, etc. The existence of uncertainty affects the reliability of risk management decision-making.
(3) Lack of effective countermeasures: Geological disasters are usually emergencies, and the requirements for real-time measures are very eager. However, effective response measures are still lack today and may lead to increased disaster losses.
(4) The cost of geological hazard risk management: Geological disaster management requires a large amount of investment, including data collection, monitoring facilities, emergency response measures, etc., which are very consuming. These costs may limit the implementation of geohazard risk management.
(5) Social awareness and participation: The management of geological disasters requires the active participation and support of all sectors of society. However, some regions may lack sufficient social awareness, which may lead to difficulties and reduced effectiveness of risk management.

## 1.3 Research Methods of Geological Disasters

Geological disaster research includes geological disaster exploration, monitoring, prediction and prevention. The purpose of geological disaster research is to deeply reveal the formation mechanism and evolution law of geological disasters, and to prevent disaster, and reduce disaster losses to a minimum. The prevention and control of geological disasters should be based on defense, supplemented by governance. To prevent geological disasters or manage geological disasters in advance, it is necessary to carry out timely and accurate spatiotemporal prediction of geological disasters. Therefore, long-term and systematic monitoring of the development of geological hazards should be arranged on the basis of exploration. Scientific analysis of a large number of reliable monitoring data will help to make the real-time and accurate predictions. Therefore, it is very important to choose appropriate monitoring methods to study corresponding geological disasters. This section will introduce from two aspects of ground survey methods and remote sensing methods.

### *1.3.1 Research Mode of Ground Equipment-Based*

#### Connotation of Ground Equipment for Geological Disaster Research

Research on geological hazards based on ground equipment mainly includes the following contents:

(1) Acquisition of geological data: Use ground equipment such as geological radar, seismograph, measuring instrument to conduct in-situ survey and data acquisition, obtain information such as geological structure, geological horizon, groundwater level, as well as data such as seismic waves and surface displacement for in-depth researches on the formation mechanism of disasters.
(2) Monitoring of geological disasters: Use ground equipment such as earthquake monitoring stations, landslide monitors, ground displacement monitors to conduct real-time monitoring of the evolution process and forecast futural trends of geological disasters.
(3) Simulation of geological processes: Use ground equipment such as physical model experiment devices and numerical simulation software to simulate and analyze geological processes, and explore the occurrence mechanism and evolution law of geological disasters.
(4) Carrying out disaster managements: Use ground equipment such as laser scanners, drones, etc. to conduct on-site surveys and disaster assessments, and provide scientific basis and technical support for disaster response and rescue work.

#### Technical System for Geological Disaster Research by Ground Equipment

Ground equipment-based technical system for geological disaster research is a multidisciplinary and comprehensively applied research system. Through the comprehensive application of various technical means, a comprehensive understanding of the geological conditions and hidden dangers of the research area will be provided for disaster prediction, prevention and emergency response. This content mainly includes the following aspects:

(1) Geological survey: Carrying out geological survey to understand the geological situation and geological structure of the study area, including stratum structure, structural form, lithology, faults and folds, so that can provide basic data for the formation and development of geological disasters.
(2) Geophysical survey: Including gravity, magnetic, electromagnetic methods to provide data support for geological disaster research by detecting the underground structure, physical and topographical characteristics of the research area.
(3) Geochemical analysis: Conducting chemical analysis of geological samples in the research area to understand the content and distribution of various elements and compounds in groundwater, surface water, soil and rocks, and provide a basis for geological disaster research.
(4) Monitoring technology: Including topographical surveying, hydrological surveying, seismic monitoring, remote sensing monitoring for the long-term and real-time monitoring of changes in topographical, hydrological, seismic and other parameters in the research area to timely discover the hidden dangers of geological disasters and provide early warning.

(5) Numerical simulation technology: Using mathematical models and computer simulation methods to numerically simulate and predict the formation and evolution of geological disasters, and provide support for decision making.

**Advantages and Disadvantages of Ground Equipment**

Research on geological hazards based on ground equipment has the following advantages and disadvantages (Guo et al., 2012; Zhang et al., 2022):

(1) Advantage: Ground equipment can directly observe geological hazards on the ground, the accuracy and reliability of the data are relatively higher. And it can provide in-situ real-time monitoring and early warning, as well as experimental verification. It can also collect various types of data, including data on geology, topography, groundwater, earthquakes, so that comprehensive geological disaster research and analysis can be carried out. In addition, ground equipment has a wide range of applications and can be applied to different types of geological disasters, such as earthquakes, landslides, ground subsidence, which has good versatility and flexibility. And comparing with other technical means such as space remote sensing technology, the cost of geological disaster research based on ground equipment is relatively lower.
(2) Disadvantages: Ground equipment-based researches can be restricted with the location and scope of study area. In many cases, it is impossible to cover the whole study site, which requires multiple equipment to cover more areas. And in inaccessible areas, and there are certain difficulties in setting the ground equipment. In addition, geological disasters usually occur in complex geological environments, such as mountains, valleys, deserts, so that environmental factors can disturb the performance of equipment. Moreover, the data obtained based on ground equipment is relatively complex with different format, size and processing approaches. The data processing process can be cumbersome and requires professional knowledge.

**Advanced Technologies of Ground Equipment-Based Geological Disaster Research**

(1) LiDAR technology

LiDAR (Light Detection and Ranging) is an active remote sensing technology that emits laser pulses to the target and uses the time delay and echo intensity of the laser pulse to obtain the three-dimensional spatial information of the target. LiDAR has the advantages of high precision, high speed, non-contact, and all-weather work, and has been widely used in various fields, such as topographic surveying, urban planning, agriculture, forestry, environmental monitoring, intelligent transportation, etc.

In geological disaster research, LiDAR can provide high-precision, high-resolution terrain data for terrain analysis, landform evolution research, and geological hazard risk assessment and prediction (Joyce et al., 2014). LiDAR can obtain high-precision 3D point cloud data on the surface for terrain analysis and landform evolution research. The analysis of topographic features and the simulation of different geomorphological processes can be used to gain a deeper understanding of the formation mechanism and development law of geological disasters. In addition, the integration of LiDAR high-precision terrain data, geological data and disaster history data, can assess the risk of geological disasters. For example, spatial analysis and simulation of landslides and debris flows can be conducted with these data, and the scope and scale of their possible occurrence can be predicted. Moreover, LiDAR can achieve real-time monitoring of the whole process of geological disasters. For example, in landslide monitoring, LiDAR can provide spatiotemporal evolution information of landslide activity through continuous measurement of surface deformation, which helps to accurately predict the development trend of landslides and take corresponding disaster prevention measures in time.

(2) Combination of ground-based sensors and Internet of Things (IoT) technology

A ground-based sensor is a device used to monitor geological hazards, which can sense changes in the ground or rocks, such as earthquakes, ground fissures, landslides, ground subsidence, and meanwhile monitor environmental factors such as temperature, humidity, wind speed. Ground-based sensors are widely used in geological exploration, seismic monitoring, groundwater level monitoring and other fields.

With the development of IoT technology, ground-based sensors are also widely used in geological hazard research. The IoT technology can achieve the interaction between ground-based sensors, real-time delivery of data, and process the data with cloud computing, so as to improve the early warning and response capabilities of geological disasters (Adeel et al., 2019; Mei et al., 2019). In addition, IoT technology can also integrate data from ground-based sensors with other data sources, such as satellite images, meteorological data, etc., to achieve comprehensive monitoring and analysis of geological disasters. Comprehensive early warning and response to geological disasters can be achieved by establishing a multi-source data-based geological disaster early warning and emergency response system.

(3) Combination of ground-based sensors and machine learning technology

As geological disasters are affected by a variety of environmental factors, it is usually difficult to accurately predict the occurrence of geological disasters solely relying on ground-based sensor data due to the hidden complex nonlinear relationships between disasters and the factors. Machine learning technology has good performance on mining the potential principles and characteristics in ground-based sensor data thereby improving the prediction accuracy and efficiency of geological disasters (Vincent et al., 2023).

In the process of combining ground-based sensors with machine learning technology for geological disaster research, ground-based sensors firstly acquire data by measuring indicators such as surface movement and structural deformation. Data cleaning, removal of outliers, feature extraction will be performed to pre-process the collected data. Afterwards, the features like displacement, velocity, acceleration will be extracted from the data to construct the dataset with the geological disaster related labels. According to the characteristics and data of geological disasters, the suitable machine learning model will be constructed, such as support vector machine, decision tree and neural network. The constructed dataset will be used to train the machine learning model, and the well-trained model can predict the probability of geological disaster occur based on new ground-based sensor data.

By comparing with the actual situation, the accuracy and reliability of the machine learning model for predicting geological disasters can be evaluated to continuously improve and optimize the prediction model.

## 1.3.2 Research Mode of Remote Sensing-Based

### Connotation of Remote Sensing Technology for Geological Disaster Research

Remote sensing technology obtains information on the earth's surface through remote sensors such as satellites and aircraft. It is widely used in the monitoring of geological disasters as it can provide advantages of high resolution, wide coverage, high efficiency, and multiple time scales, helping researchers to study the dynamic process and spatial distribution of geological disasters, and providing scientific basis for preventing, mitigating, and responding to geological disasters. Remote sensing technology can be applied in the study of geological disasters from following aspects:

(1) Topography and geomorphology analysis: Digital elevation model (DEM) data obtained from remote sensing can be analyzed to understand the topography and geomorphology background of geological disasters.
(2) Landcover classification: Through the classification and interpretation of satellite remote sensing data, different land use types are obtained to study the spatial distribution of geological disasters.
(3) Geological structure analysis: Remote sensing data such as DEM, surface reflectance and spectral indices can be used to analyze the situation of underground structures and study the causes of geological disasters.
(4) Surface deformation monitoring: Surface deformation can be monitored with remote sensing data obtained in different dates, so that the mechanism of geological disasters such as earthquakes and landslides can be studied.
(5) Hydrogeological analysis: Remote sensing technology can be used to obtain hydrological information, such as the location, size, and velocity of water bodies such as rivers and lakes, so as to study the causes and laws of floods, mountain torrents, and debris flows.

### Technical System of Remote Sensing Technology for Geological Disaster Research

Remote sensing technology can provide multi-source, multi-angle, and multi-temporal data, providing a broad space and time scale for geological disaster research, and can be applied to geological disaster early warning, monitoring, evaluation and prevention. The technical method system of remote sensing technology for geological disaster research mainly includes the following aspects:

(1) Image processing: Perform preprocessing, atmospheric correction, geometric correction, image enhancement and other processing to improve the quality and usability of images.
(2) Disaster classification and identification: According to the characteristics of geological disasters, feature extraction and classification algorithms are used to achieve automatic identification of different types of geological disasters.
(3) Change detection and monitoring: By the difference comparison and change detection of remote sensing images in different time periods, the monitoring and early warning of geological disasters can be achieved.
(4) Spatial analysis and evaluation: Use GIS related tools to conduct spatial analysis and evaluation on remote sensing images, calculate the disaster area, disaster distribution, loss assessment and other indicators in the disaster area, and provide scientific basis for decision-making.
(5) Simulation and prediction: By constructing the geological disaster prediction model, using remote sensing images and other relevant data, the disaster probability prediction and disaster impact assessment can be conducted.

**Advantages and Disadvantages of Remote Sensing Technology**

The advantages and disadvantages of remote sensing technology for geological disaster research are as follow (Qu et al., 2020; Ren et al., 2019):

(1) Advantage: Remote sensing technology can cover a wide area and obtain a large amount of data, and can quickly obtain information on surface morphology and landform changes before and after geological disasters occur in a short period of time. The high-resolution remote sensing images can capture some small surface changes, such as ground fissures and deformation of landslides. The images in different periods can achieve long-term regional monitoring of geological disasters. In addition, the remote sensing image data can be processed automatically by computer, which can greatly reduce labor and time costs.
(2) Disadvantage: Remote sensing technology can only obtain surface information, and cannot directly obtain underground geological information and geological structures, so it needs to cooperate with other geological survey methods for comprehensive research. And remote sensing images have a large amount of data that requires advanced computing devices which have not been popularized in most regions. Moreover, remote sensing technology requires clear weather with low cloud cover, otherwise it will affect the acquisition and quality of data. And there are inherent errors in the data obtained by remote sensing technology, and accuracy verification and correction are required to ensure the accuracy of the data.

**Advanced Remote Sensing Technology for Geological Hazard Research**

(1) High resolution optical remote sensing

High-resolution optical data refers to the optical image data of the earth's surface obtained by high-resolution satellites, aircraft or drones. These data usually include high-resolution color or multispectral images, which can be used to identify and analyze the shape, structure, texture and other information of surface objects.

High-resolution optical remote sensing has important application value in geological hazard research, which can provide support for disaster monitoring, assessment, early warning, emergency response, and promote the development of geological disaster research and the practice of disaster prevention and control. In geological hazard research, high-resolution optical remote sensing can provide a series of detail information (Amatya et al., 2019). It can extract surface topography and landform information, such as mountains, canyons, rivers, lakes for studying the formation and evolution process of geological hazards. The distribution and state of vegetation also have a great influence on the occurrence and evolution of geological disasters. High-resolution optical remote data can provide vegetation information, such as vegetation coverage, type, height. The structure and texture of soil are also related to geological disasters like landslide and debris flow. High-resolution optical data can be used to extract soil information, such as soil type, thickness, color, etc. In addition, high-resolution optical remote sensing can be used to extract information about human factors, such as buildings, roads, farmland, etc., which is helpful for studying the impact of human activities on geological hazards. Using multi-temporal image data for differential interferometry can also accurately monitor small surface deformations, such as land subsidence and ground fissures, so as to better study the formation and evolution of geological disasters.

(2) Aerial LiDAR

Aerial Light Detection and Ranging (LiDAR) is a remote sensing technology that uses laser beams to measure distances between an airborne platform and the ground surface. The principle is same to the ground LiDAR. Aerial LiDAR is often used for a variety of applications, including surveying, mapping, urban planning, forestry, and environmental monitoring. It can be used to create high-resolution digital elevation models (DEMs).

LiDAR data has been widely used in geological hazard research as it can helping researchers better understand the formation mechanism and evolution of geological hazards (She et al., 2021). Specifically, LiDAR can quickly and accurately obtain terrain information, including terrain elevation, slope angle, slope aspect, terrain curvature. These data can be used to generate the high-resolution 3D landform model, which can help researchers identify and analyze geological disaster risk areas. For example, the 3D model can help assess the scale and scope of the disaster after the earthquake, and optimize rescue and recovery work. In addition, LiDAR can perform high-precision measurement and monitoring of landslides, including landslide shape, deformation, movement velocity, which is helpful for studying the movement mechanism of landslides and predicting the risk of it.

(3) InSAR

Interferometric Synthetic Aperture Radar (InSAR) is a technology that uses satellite synthetic aperture radar (SAR) to monitor surface deformation. This technology can make multiple observations of the same area by satellite, and use the phase change of radar wave propagation to calculate the deformation of the surface. InSAR data can provide high-precision, high-resolution surface deformation information, which has advantages that remote sensing data do not have.

InSAR data are widely used in geological hazard research. Geological hazards include, but are not limited to, earthquakes, volcanoes, landslides, ground subsidence, and glacial dynamics. In earthquake research, InSAR data can be used to measure surface deformation and source parameters caused by earthquakes (Zheng et al., 2022). In volcano research, it can detect surface deformation before and after volcanic eruptions. And in landslide research, InSAR data can measure the sliding and deformation of the surface. In the study of ground subsidence, InSAR data can be used to monitor the situation of surface subsidence and provide reference for urban planning and land use. And for glacier activity, InSAR data can be used to monitor the morphological changes and its speed of glaciers, and predict the trend and scale of glacier melting.

## 1.4 Conclusions and Future Directions

### *1.4.1 Key Findings and Insights from the Review*

Research on geological disasters has a very long history. Over the past few decades, research on geological hazards has yielded many important discoveries and insights.

Earthquake research mainly focused on the physical mechanism, prediction and disaster prevention and mitigation of earthquakes (Chiang et al., 2022; Dong & Luo, 2022; Wang et al., 2017; Wu et al., 2021b). Advances in seismic monitoring technology have allowed scientists to detect and measure smaller earthquakes more accurately than ever before. This has led to a better understanding of earthquake patterns and the ability to detect potential earthquakes before they occur. Scientists have made progress in understanding the different types of faults that cause earthquakes and the types of tectonic activity that lead to seismic activity (Ali & Abdelrahman, 2022). This has allowed for more accurate predictions of earthquake hazards and improved building codes in earthquake-prone areas. Machine learning models are now able to simulate earthquakes and predict their effects with greater accuracy (Beroza et al., 2021). This has allowed for more informed emergency planning and better disaster response strategies. However, there is still much to be learned about earthquakes, and ongoing research is necessary to continue to improve our understanding of these complex phenomena.

Recent research in the field of debris flows has focused on understanding the triggering mechanisms of debris flows and the factors that contribute to their mobility and destructive potential (Kang & Kim, 2016; Kasim et al., 2016). Advances in remote sensing technologies such as satellite imagery and LiDAR and processing methods like deep learning, have enabled researchers to monitor and predict debris flow events with greater accuracy. This includes developing early warning systems that can provide timely alerts to the at-risk communities (Zhao et al., 2022). In addition, researchers are also studying the effects of rainfall and other environmental factors on the hydrology of debris flows, including the flow velocity and sediment transport capacity (Syarifuddin et al., 2017). This information can help to better understand and predict the behavior of debris flows.

In landslide study, a number of researchers are studying the various factors that can trigger landslides, including rainfall, seismic activity, slope geometry, and geology (Cui et al., 2022b; Dikshit et al., 2020; Tsai et al., 2019; Wasowski et al., 2011). In addition, understanding the dynamics of landslides is critical to predicting their behavior and mitigating their impact. Recent research has focused on the movement of landslides, including the velocity, direction, and type of movement, as well as the factors that contribute to their mobility (Li et al., 2017; Squarzoni et al., 2020). Susceptibility of landslide is a hot topic with the development of remote sensing, GIS technologies, and artificial intelligence (Azarafza et al., 2021). The landslide susceptibility maps can help identify areas that are most susceptible to landslides, which can be used to inform land use planning and hazard mitigation strategies. For landslide mitigation, researchers are developing more new strategies, including the use of barriers and other structures to divert or contain landslide movement (Prastica et al., 2019), and the development of evacuation plans to help protect at-risk communities (Iskandar et al., 2021).

In ground instability study, advanced satellite and aerial remote sensing technologies have enabled researchers to monitor and predict ground subsidence, ground fissure and sinkhole with greater accuracy (Kim et al., 2019; Sorkhabi et al., 2022). With this, researchers are studying the various factors that contribute to ground instability, including natural processes such as soil compaction and subsurface erosion, as well as human activities such as groundwater extraction and land use change (Jia et al., 2021; Othman & Abotalib, 2019). For its prevention and mitigation, soil reinforcement techniques, injection of grout or other materials to stabilize the soil, and land use planning strategies are recently researched to reduce the risk of ground instability (Li & Su, 2022; Park et al., 2020).

Overall, the research on geological hazards in the past few decades has been continuously deepened and expanded, from physical mechanisms to disaster prevention and mitigation, from a single type to multiple types, from a single region to the global scale, forming a relatively complete research system and theory frame. These research results not only enrich people's understanding of the earth's natural environment, but also provide scientific basis and technical support for disaster prevention and mitigation and social development.

### 1.4.2 Gaps and Challenges in Current State of Geological Disaster Research and Management

Geological disasters have caused significant losses in human society, so the research and management of geological disasters are very important. There are still following challenges in current geological disaster research and management (Cui et al., 2021). Some of these challenges include:

(1) Limited ground truth data: Ground survey is still necessary in geological disaster study today for the validation of remote sensing-based researches. Collecting data on geological hazards can be difficult and expensive, particularly in remote or inaccessible areas, like regions with high elevation, slope angle, and dense vegetation coverage (Fan et al., 2017), so the ability to accurately predict and mitigate their impact may be compromised.
(2) Uncertainty in hazard assessments: Geological disasters are complex and can be influenced by a wide range of factors, making hazard assessments challenging. Different uncertainties affect the assessing accuracy and can lead to difficulties to make decisions about risk management and mitigation strategies (Khalaj et al., 2020; Lee, 2016). Therefore, advanced uncertainty quantification methods for various geological disasters are needed.
(3) Changing hazard patterns: The patterns and frequency of geological hazards can change over time and study area, making it difficult to accurately predict their occurrence. For example, climate change is likely to have an impact on the local occurrence and severity of some geological hazards, such as landslides and flooding, but the affects can be various in different study sites (Kirschbaum et al., 2020).
(4) Technological innovation: With the continuous advancement of science and technology, the research and management of geological disasters are facing the pressure of technological innovation. New technical means, new data collection and processing methods like advanced sensors and state of the art algorithms need to be continuously explored and applied in order to better improve the level of geological disaster research and management (Cui et al., 2022a, 2022b).

For the disaster management, there are following challenges:

(1) Inadequate risk communication: Communicating risk to the public and other stakeholders is critical in ensuring that appropriate mitigation measures are taken. However, there can be a disconnect between the scientific understanding of hazards and the public's perception of risk (Yamori, 2020).
(2) Population growth: With the growth of population and the acceleration of urbanization, the interaction between human activities and geological environment is increasing, and the risk of geological disasters is also increasing (Zhang & Li, 2020).
(3) Poverty and resource shortage: The prevention and control of geological disasters requires a lot of human, material and financial support, like effective enhanced materials to prevent the extension of ground fissure and subsidence,

which are limited for regions suffer from poverty (Xiang et al., 2023). In addition, the coupling effects of resource shortage and geological disasters have not been well understood, unsuitable financial support policies may lead to more problems (Lan et al., 2022).

(4) Imperfect management system: Geological disaster management requires cross-departmental and cross-industry collaboration, but the current management system is of a lot of problems, such as poor coordination and unclear responsibilities among various departments, which also limit geological disaster management (Yi et al., 2012).

### *1.4.3 Future Directions and Opportunities for Advancing Understanding and Addressing Geological Disasters*

To better understand and address geohazards, future directions and opportunities may include the following:

(1) Strengthen research on geological disaster monitoring and early warning technology. The occurrence of geological disasters is often uncertain and sudden, so timely early warning and prediction technology is very important. Real-time monitoring and early warning of geological disasters can be achieved by the integration of satellite remote sensing, geological radar, IoT disaster monitoring and other technical devices, and the ability of disaster prevention and mitigation can be improved (Wang et al., 2022).

(2) Improve the precision and accuracy of geological disaster risk assessment. Geological disaster risk assessment is an important part of disaster reduction work. Technical means such as GIS and digital earth can be used to conduct refined assessment and prediction of disaster risk, and provide scientific basis for government decision-making and public disaster prevention and reduction (Shi et al., 2020).

(3) Explore new technologies and models for geological disaster management. Traditional geological disaster management methods are often based on civil engineering methods. New technologies such as ecological restoration, artificial intelligence, and machine learning can be used in combination with new models such as social participation and technological innovation to achieve sustainable development of geological disaster management (Cui et al., 2021).

(4) Establish a global geological disaster response system. Geological disasters are a global problem. In the future, through international cooperation and exchanges, a global geological disaster response system can be established to jointly promote disaster prevention, mitigation and post-disaster recovery, so as to ensure the safety and sustainable development of human society (Andrews & Quintana, 2015).

Overall, the directions and opportunities for future geological disaster monitoring and prevention are diverse. With technological innovation, system construction and international cooperation, we will continuously improve the ability of disaster prevention and mitigation, and achieve effective governance of geological disasters and sustainable development.

## References

Adeel, A., Gogate, M., Farooq, S., Ieracitano, C., Dashtipour, K., Larijani, H., & Hussain, A. (2019). A survey on the role of wireless sensor networks and IoT in disaster management. In *Geological disaster monitoring based on sensor networks* (pp. 57–66).

Ali, S. M., & Abdelrahman, K. (2022). Earthquake occurrences of the major tectonic terranes for the Arabian Shield and their seismic hazard implications. *Frontiers in Earth Science, 10*, 851737.

Amatya, P., Kirschbaum, D., & Stanley, T. (2019). Use of very high-resolution optical data for landslide mapping and susceptibility analysis along the Karnali highway, Nepal. *Remote Sensing, 11*(19), 2284.

Andrews, R. J., & Quintana, L. M. (2015). Unpredictable, unpreventable and impersonal medicine: Global disaster response in the 21st century. *EPMA Journal, 6*, 1–12.

Azarafza, M., Azarafza, M., Akgün, H., Atkinson, P. M., & Derakhshani, R. (2021). Deep learning-based landslide susceptibility mapping. *Scientific Reports, 11*(1), 24112.

Bagheri-Gavkosh, M., Hosseini, S. M., Ataie-Ashtiani, B., Sohani, Y., Ebrahimian, H., Morovat, F., & Ashrafi, S. (2021). Land subsidence: A global challenge. *Science of the Total Environment, 778*, 146193.

Beroza, G. C., Segou, M., & Mostafa Mousavi, S. (2021). Machine learning and earthquake forecasting—Next steps. *Nature Communications, 12*(1), 4761.

Beven, K. J., Almeida, S., Aspinall, W. P., Bates, P. D., Blazkova, S., Borgomeo, E., Freer, J., Goda, K., Hall, J. W., Phillips, J. C., & Simpson, M. (2018). Epistemic uncertainties and natural hazard risk assessment—Part 1: A review of different natural hazard areas. *Natural Hazards and Earth System Sciences, 18*(10), 2741–2768.

Bird, P., & Liu, Z. (2007). Seismic hazard inferred from tectonics: California. *Seismological Research Letters, 78*(1), 37–48.

Caramanna, G., Ciotoli, G., & Nisio, S. (2008). A review of natural sinkhole phenomena in Italian plain areas. *Natural Hazards, 45*(2), 145–172.

Causes, L. (2001). *Landslide types and processes*. US Geological Survey.

Chae, B.-G., Park, H.-J., Catani, F., Simoni, A., & Berti, M. (2017). Landslide prediction, monitoring and early warning: A concise review of state-of-the-art. *Geosciences Journal, 21*, 1033–1070.

Chen, F., Lin, H., Zhang, Y., & Lu, Z. (2012). Ground subsidence geo-hazards induced by rapid urbanization: Implications from InSAR observation and geological analysis. *Natural Hazards and Earth System Sciences, 12*(4), 935–942.

Chen, J., Li, J., Qin, X., Dong, Q., & Sun, Y. (2009). RS and GIS-based statistical analysis of secondary geological disasters after the 2008 Wenchuan earthquake. *Acta Geologica Sinica-English Edition, 83*(4), 776–785.

Chen, Y., Lin, H., & Liu, B. (2022). Review of research progresses and application of geothermal disaster prevention on large-buried tunnels. *Applied Sciences, 12*(21), 10950.

Chiang, Y.-J., Chin, T.-L., & Chen, D.-Y. (2022). Neural network-based strong motion prediction for on-site earthquake early warning. *Sensors, 22*(3), 704.

Cui, P., Ge, Y., Li, S., Li, Z., Xu, X., Zhou, G. G., Chen, H., Wang, H., Lei, Y., Zhou, L., & Yi, S. (2022a). Scientific challenges in disaster risk reduction for the Sichuan-Tibet railway. *Engineering Geology, 309*, 106837.

Cui, S., Wu, H., Pei, X., Yang, Q., Huang, R., & Guo, B. (2022b). Characterizing the spatial distribution, frequency, geomorphological and geological controls on landslides triggered by the 1933 Mw 7.3 Diexi earthquake, Sichuan, China. *Geomorphology, 403*, 108177.

Cui, P., Peng, J., Shi, P., Tang, H., Ouyang, C., Zou, Q., Liu, L., Li, C., & Lei, Y. (2021). Scientific challenges of research on natural hazards and disaster risk. *Geography and Sustainability, 2*(3), 216–223.

Da-yu, G., & Li, Z. (2000). Ground fissure hazards in USA and China. *Acta Seismologica Sinica, 13*(4), 466–476.

De Luca, D. L., & Versace, P. (2017). Diversity of rainfall thresholds for early warning of hydro-geological disasters. *Advances in Geosciences, 44*, 53–60.

Dhakal, S. (2015). *Disasters in Nepal*. Tribhuvan University, Central Department of Environmental Science Kirtippur, Nepal and United Nations Development Programme.

Dikshit, A., Sarkar, R., Pradhan, B., Segoni, S., & Alamri, A. M. (2020). Rainfall induced landslide studies in Indian Himalayan region: A critical review. *Applied Sciences, 10*(7), 2466.

Dong, L., Cao, J., & Liu, X. (2022). Recent developments in sea-level rise and its related geological disasters mitigation: A review. *Journal of Marine Science and Engineering, 10*(3), 355.

Dong, L., & Luo, Q. (2022). Investigations and new insights on earthquake mechanics from fault slip experiments. *Earth-Science Reviews, 228*, 104019.

Fan, X., Xu, Q., Scaringi, G., Dai, L., Li, W., Dong, X., Zhu, X., Pei, X., Dai, K., & Havenith, H. B. (2017). Failure mechanism and kinematics of the deadly June 24th 2017 Xinmo landslide, Maoxian, Sichuan, China. *Landslides, 14*, 2129–2146.

Fanos, A. M., & Pradhan, B. (2018). Laser scanning systems and techniques in rockfall source identification and risk assessment: A critical review. *Earth Systems and Environment, 2*(2), 163–182.

Guo, H., Liu, L., Fan, X., Li, X., & Zhang, L. (2012). *Earthquake research and analysis-statistical studies, observations and planning*. INTECH Open Access Publisher.

Hürlimann, M., Rickenmann, D., Medina, V., & Bateman, A. (2008). Evaluation of approaches to calculate debris-flow parameters for hazard assessment. *Engineering Geology, 102*, 152–163.

Iskandar, I., Andika, T., & Wulandari, W. (2021). The model of nonstructural mitigation policy to the landslide prone residential areas in Lebong, Bengkulu. *Yuridika, 36*(2), 333–348.

Jia, Z., Qiao, J., Peng, J., Lu, Q., Xia, Y., Zang, M., Wang, F., & Zhao, J. (2021). Formation of ground fissures with synsedimentary characteristics: A case study in the Linfen Basin, northern China. *Journal of Asian Earth Sciences, 214*, 104790.

Jiang, W., Rao, P., Cao, R., Tang, Z., & Chen, K. (2017). Comparative evaluation of geological disaster susceptibility using multi-regression methods and spatial accuracy validation. *Journal of Geographical Sciences, 27*, 439–462.

Joyce, K. E., Samsonov, S. V., Levick, S. R., Engelbrecht, J., & Engelbrecht, S. (2014). Mapping and monitoring geological hazards using optical, LiDAR, and synthetic aperture RADAR image data. *Natural Hazards, 73*, 137–163.

Kang, H.-S., & Kim, Y.-T. (2016). The physical vulnerability of different types of building structure to debris flow events. *Natural Hazards, 80*, 1475–1493.

Kasim, N., Taib, K. A., Mukhlisin, M., & Kasa, A. (2016). Triggering mechanism and characteristic of debris flow in Peninsular Malaysia. *American Journal of Engineering Research, 5*, 112–119.

Khalaj, S., BahooToroody, F., Abaei, M. M., BahooToroody, A., De Carlo, F., & Abbassi, R. (2020). A methodology for uncertainty analysis of landslides triggered by an earthquake. *Computers and Geotechnics, 117*, 103262.

Kim, J.-W., Lu, Z., & Kaufmann, J. (2019). Evolution of sinkholes over Wink, Texas, observed by high-resolution optical and SAR imagery. *Remote Sensing of Environment, 222*, 119–132.

Kirschbaum, D., Kapnick, S. B., Stanley, T., & Pascale, S. (2020). Changes in extreme precipitation and landslides over High Mountain Asia. *Geophysical Research Letters, 47*(4), e2019GL085347.

Kusky, T. M. (2003). *Geological hazards: A sourcebook*. Greenwood Publishing Group.

Kusky, T. M. (2008). *Volcanoes: Eruptions and other volcanic hazards*. Infobase Publishing.

Lan, H., Tian, N., Li, L., Liu, H., Peng, J., Cui, P., Zhou, C., Macciotta, R., & Clague, J. J. (2022). Poverty control policy may affect the transition of geological disaster risk in China. *Humanities and Social Sciences Communications, 9*(1), 1–7.

Lee, E. M. (2016). Landslide risk assessment: The challenge of communicating uncertainty to decision-makers. *Quarterly Journal of Engineering Geology and Hydrogeology, 49*(1), 21–35.

Li, P., & Su, F. (2022). Unidirectional geosynthetic reinforcement design for bridging localized sinkholes in transport embankments. *Mathematical Problems in Engineering.*

Li, Y., Wang, X., & Mao, H. (2020). Influence of human activity on landslide susceptibility development in the Three Gorges area. *Natural Hazards, 104*, 2115–2151.

Li, Z., Huang, X., Xu, Q., Yu, D., Fan, J., & Qiao, X. (2017). Dynamics of the Wulong landslide revealed by broadband seismic records. *Earth, Planets and Space, 69*, 1–10.

Li, Z., Zhou, F., Han, X., Chen, J., Li, Y., Zhai, S., Han, M., & Bao, Y. (2021). Numerical simulation and analysis of a geological disaster chain in the Peilong valley, SE Tibetan Plateau. *Bulletin of Engineering Geology and the Environment, 80*, 3405–3422.

Liu, M., He, Y., Wang, J., Lee, H. P., & Liang, Y. (2015). Hybrid intelligent algorithm and its application in geological hazard risk assessment. *Neurocomputing, 149*, 847–853.

Mei, G., Xu, N., Qin, J., Wang, B., & Qi, P. (2019). A survey of Internet of Things (IoT) for geohazard prevention: Applications, technologies, and challenges. *IEEE Internet of Things Journal, 7*(5), 4371–4386.

Menon, N. V. (2012). Challenges in disaster management. *Yojana, 13*, 13–16.

Nathe, S. K. (2000). Public education for earthquake hazards. *Natural Hazards Review, 1*(4), 191–196.

Othman, A., & Abotalib, A. Z. (2019). Land subsidence triggered by groundwater withdrawal under hyper-arid conditions: Case study from Central Saudi Arabia. *Environmental Earth Sciences, 78*, 1–8.

Pardeshi, S. D., Autade, S. E., & Pardeshi, S. S. (2013). Landslide hazard assessment: Recent trends and techniques. *SpringerPlus, 2*, 1–11.

Park, J., Chung, Y., & Hong, G. (2020). Reinforcement effect of a concrete mat to prevent ground collapses due to buried pipe damage. *Applied Sciences, 10*(16), 5439.

Prastica, R. M., Apriatresnayanto, R., & Marthanty, D. R. (2019). Structural and green infrastructure mitigation alternatives prevent Ciliwung River from water-related landslide. *International Journal on Advanced Science, Engineering and Information Technology, 9*(6), 1825–1832.

Qu, T., Tian, G., & Tang, M. (2020). Research on methods and application of remote sensing technology in geological hazard monitoring and management. *IOP Conference Series: Earth and Environmental Science, 514*(2), 022075.

Ren, H., Zhao, Y., Xiao, W., & Hu, Z. (2019). A review of UAV monitoring in mining areas: Current status and future perspectives. *International Journal of Coal Science & Technology, 6*, 320–333.

She, J., Mabi, A., Liu, Z., Sheng, M., Dong, X., Liu, F., & Wang, S. (2021). Analysis using high-precision airborne LiDAR data to survey potential collapse geological hazards. *Advances in Civil Engineering*, 1–10.

Shi, P., Ye, T., Wang, Y., Zhou, T., Xu, W., Du, J., Wang, J. A., Li, N., Huang, C., Liu, L., & Chen, B. (2020). Disaster risk science: A geographical perspective and a research framework. *International Journal of Disaster Risk Science, 11*, 426–440.

Sorkhabi, O. M., Kurdpour, I., & Sarteshnizi, R. E. (2022). Land subsidence and groundwater storage investigation with multi sensor and extended Kalman filter. *Groundwater for Sustainable Development, 19*, 100859.

Squarzoni, G., Bayer, B., Franceschini, S., & Simoni, A. (2020). Pre- and post-failure dynamics of landslides in the Northern Apennines revealed by space-borne synthetic aperture radar interferometry (InSAR). *Geomorphology, 369*, 107353.

Sun, R., Gao, G., Gong, Z., & Wu, J. (2020). A review of risk analysis methods for natural disasters. *Natural Hazards, 100*(2), 571–593.

Sun, X., Yu, C., Li, Y., & Rene, N. N. (2022). Susceptibility mapping of typical geological hazards in Helong City affected by volcanic activity of Changbai Mountain, Northeastern China. *ISPRS International Journal of Geo-Information, 11*(6), 344.

Syarifuddin, M., Oishi, S., Legono, D., Hapsari, R. I., & Iguchi, M. (2017). Integrating X-MP radar data to estimate rainfall induced debris flow in the Merapi volcanic area. *Advances in Water Resources, 110*, 249–262.

Takahashi, T. (1981). Debris flow. *Annual Review of Fluid Mechanics, 13*(1), 57–77.

Theron, A., & Engelbrecht, J. (2018). The role of earth observation, with a focus on SAR interferometry, for sinkhole hazard assessment. *Remote Sensing, 10*(10), 1506.

Tsai, H.-Y., Tsai, C.-C., & Chang, W.-C. (2019). Slope unit-based approach for assessing regional seismic landslide displacement for deep and shallow failure. *Engineering Geology, 248*, 124–139.

Vandewater, C. J., Dunne, W. M., Mauldon, M., Drumm, E. C., & Bateman, V. (2005). Classifying and assessing the geologic contribution to rockfall hazard. *Environmental & Engineering Geoscience, 11*(2), 141–154.

Vincent, S., Pathan, S., & Benitez, S. R. (2023). Machine learning based landslide susceptibility mapping models and GB-SAR based landslide deformation monitoring systems: Growth and evolution. *Remote Sensing Applications: Society and Environment*, 100905.

Wang, H. B., Wu, S. R., Shi, J. R., & Li, B. (2013). Qualitative hazard and risk assessment of landslides: A practical framework for a case study in China. *Natural Hazards, 69*, 1281–1294.

Wang, Q., Guo, Y., Yu, L., & Li, P. (2017). Earthquake prediction based on spatio-temporal data mining: An LSTM network approach. *IEEE Transactions on Emerging Topics in Computing, 8*(1), 148–158.

Wang, S., Chen, R., Huang, Z., Jing, H., & Wang, X. (2022). Beidou+ PS-InSAR technology for substation geological subsidence monitoring and early warning research. In *IEEE 2022 5th World Conference on Mechanical Engineering and Intelligent Manufacturing (WCMEIM)* (pp. 456–461).

Wang, X., Zhang, C., Wang, C., Liu, G., & Wang, H. (2021a). GIS-based for prediction and prevention of environmental geological disaster susceptibility: From a perspective of sustainable development. *Ecotoxicology and Environmental Safety, 226*, 112881.

Wang, Y., Wen, H., Sun, D., & Li, Y. (2021b). Quantitative assessment of landslide risk based on susceptibility mapping using random forest and GeoDetector. *Remote Sensing, 13*(13), 2625.

Wasowski, J., Keefer, D. K., & Lee, C.-T. (2011). Toward the next generation of research on earthquake-induced landslides: Current issues and future challenges. *Engineering Geology, 122*(1–2), 1–8.

Wu, C., Guo, Y., & Su, L. (2021a). Risk assessment of geological disasters in Nyingchi, Tibet. *Open Geosciences, 13*(1), 219–232.

Wu, Z., Ma, T., Jiang, H., & Jiang, C. (2021b). Multi-scale seismic hazard and risk in the China mainland with implication for the preparedness, mitigation, and management of earthquake disasters: An overview. *International Journal of Disaster Risk Reduction, 4*, 21–33.

Xiang, M., Yang, J., Han, S., Liu, Y., Wang, C., & Wei, F. (2023). Spatial coupling relationship between multidimensional poverty and the risk of geological disaster. *Local Environment*, 1–19.

Xu, L., Meng, X., & Xu, X. (2014). Natural hazard chain research in China: A review. *Natural Hazards, 70*, 1631–1659.

Yamori, K. (2020). *Disaster risk communication*. Springer Singapore.

Yi, L., Ge, L., Zhao, D., Zhou, J., & Gao, Z. (2012). An analysis on disasters management system in China. *Natural Hazards, 2012*(60), 295–309.

Zhang, F., Pei, H., Zhu, H., & Wang, L. (2021). Research review of large deformation monitoring of rock and soil. *IOP Conference Series: Earth and Environmental Science, 861*(4), 042030.

Zhang, Y., Ma, X., Quan, H., Zeng, H., & Chen, Y. (2022). Research on environmental geophysical methods in geological hazards monitoring. *International Journal of Environmental Protection and Policy, 10*(4), 92.

Zhang, Z., & Li, Y. (2020). Coupling coordination and spatiotemporal dynamic evolution between urbanization and geological hazards—A case study from China. *Science of the Total Environment, 728*, 138825.

Zhao, Y., Meng, X., Qi, T., Li, Y., Chen, G., Yue, D., & Qing, F. (2022). AI-based rainfall prediction model for debris flows. *Engineering Geology, 296*, 106456.

Zheng, Z., Xie, C., He, Y., Zhu, M., Huang, W., & Shao, T. (2022). Monitoring potential geological hazards with different InSAR algorithms: The case of western Sichuan. *Remote Sensing, 14*(9), 2049.

Zhou, H.-J., Wang, X., & Yuan, Y. (2015). Risk assessment of disaster chain: Experience from Wenchuan earthquake-induced landslides in China. *Journal of Mountain Science, 12*, 1169–1180.

# Chapter 2
# Principles and Methods of Intelligent Interpretation of Geological Disasters

**Abstract** Intelligent interpretation of geological disasters means to use new technologies and methods such as new-generation artificial intelligence, especially deep learning technology, to achieve automatic identification, prediction, and susceptibility assessment of geological disasters. Application of artificial intelligence technology to automate the analysis and interpretation of geological disasters can greatly improve the efficiency of data processing, shorten the response time, and provide accurate disaster information in time. This chapter will state the principle and methods to achieve the intelligent interpretation by introducing the ability and procedure of deep learning working on remote sensing image and the state-of-the-art deep learning architectures that are widely used in geological disaster researches.

## 2.1 Principles of Intelligent Interpretation of Geological Disasters

### *2.1.1 Ability of Deep Learning in Feature Extraction of Remote Sensing Images*

Remote sensing images usually have high spatial and spectral resolutions, and deep learning can effectively learn and extract features from these images, which are difficult for traditional feature extraction methods (Romero et al., 2015). The multiple neural network layers of deep learning models can autonomously learn abstract features from data and provide efficient and accurate method for feature extraction of remote sensing images. Different deep learning networks adapt different feature extraction strategies, can mainly categorized to convolutional neural network (CNN), recurrent neural network (RNN) and deep generative model (DGM). CNN-based models can extract spatial features from the input images by using convolutional layers to detect patterns and edges in the images (Xie et al., 2019). The output of the convolutional layers is then fed into fully connected layers for classification or regression. RNN is particularly useful for processing sequential data. It uses feedback loops to pass information from one step of the sequence to the next, allowing them

 
W. Chen et al., *Intelligent Interpretation for Geological Disasters*,
https://doi.org/10.1007/978-981-99-5822-1_2

to model temporal dependencies in the data (Ma et al., 2020). DGM, such as autoencoder, is a type of neural network that is commonly used for unsupervised learning. In the context of remote sensing images, autoencoder can be used to extract features from the input images by encoding them into a lower-dimensional latent space and then decoding them back to the original image (Lv et al., 2018). The process of encoding and decoding can be used to learn a compressed representation of the input image. Once the autoencoder has been trained, the encoder part of the network can be used as a feature extractor for downstream tasks such as classification.

**Superiority of Deep Learning in Feature Extraction of Remote Sensing Images**

Comparing with the traditional handcrafted method, the feature extraction of deep learning has the following advantages (Chen et al., 2016; Jogin et al., 2018; Liang et al., 2017; Mayer & Jacobsen, 2020):

(1) Automatic feature extraction: Deep learning algorithms can automatically extract features from the raw data without the need for human intervention.
(2) Ability to process complex data: Remote sensing images have complex spatial structures and multi-angle features. Deep learning algorithms can learn and recognize patterns from large datasets with high dimensionality without the need for explicit feature extraction.
(3) High accuracy: Deep learning models can extract features with high accuracy and efficiency, which is difficult to achieve with manual feature extraction methods.
(4) Strong scalability: Deep learning can handle large datasets and increase computational resources to accommodate growing volumes of data based on parallel processing and distributed computing.
(5) Robustness to noise and variability: Deep learning models can handle noisy and variable data better than traditional methods, as they can learn to filter out irrelevant information and focus on the most important features.

**Problems of Deep Learning in Feature Extraction of Remote Sensing Images**

Although deep learning has many advantages in feature extraction of remote sensing images, its inherent problems should be noticed and minimized in practical applications (Abdullah et al., 2019; Kakogeorgiou & Karantzalos, 2021; Ying, 2019):

(1) Preprocessing: Preprocessing of remote sensing data can be challenging due to its complexity and heterogeneity, which can make it difficult to optimize preprocessing pipelines for deep learning models.
(2) Limited labeled data: Deep learning models require a large amount of labeled data for training, but obtaining such data can be extremely expensive and time-consuming in the field of remote sensing.
(3) Overfitting: Overfitting can occur when the model is trained on a limited dataset and results in poor generalization to new data. Due to the diversity and complexity of remote sensing data, deep learning models are prone to the overfitting problems.

(4) Poor interpretability: Deep learning models are often complex and difficult to interpret, which makes it challenging to understand how they work and what features they extract.
(5) Computational resources: Training deep learning models requires significant computational resources, such as GPUs and large amounts of memory. This can be a barrier to researchers and organizations with limited resources.

### *2.1.2 Recognizability of Key Features or Patterns of Geological Disasters Based on Deep Learning*

With abovementioned feature extraction strategies, deep learning can recognize key features geological disasters from various data sources such as remote sensing imagery, DEM, geological maps, and seismic data.

(1) Landslide, debris flow, rockfall, collapse

The primary key features for the recognition of landslides, rockfall and collapse are terrain slope and curvature (Lee & Talib, 2005; Li et al., 2020b; Shirzadi et al., 2012). These geological disasters typically occur on steep slopes, and the curvature can be used to identify the location of ridges and valleys where landslide or rockfall may occur. These topographical features can be obtained from DEM image data (Fig. 2.1) and CNNs can be trained to recognize the topographic signatures of these geological disasters from raw DEM data, such as ridges, scarps, or toe areas.

Vegetation cover can be used to identify areas of high landslide risk, as vegetated slopes are less likely to experience landslides than bare slopes (Glade, 2003). Vegetation related information are usually indicated by using bands or spectral vegetation index obtained from optical remote sensing. With this image data (Fig. 2.2), CNN can recognize the vegetation categories and density of vegetation cover that account for the landslide susceptibility.

In addition, water also plays a key role in landslides, and information on soil moisture, precipitation, and groundwater levels can be used to predict landslide or

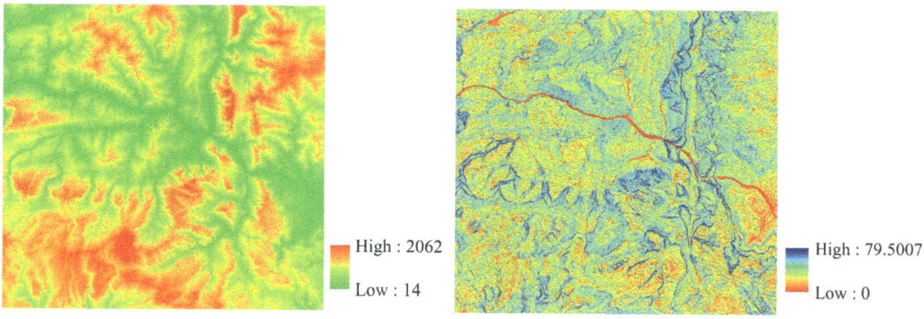


**Fig. 2.1** Terrain features extracted from DEM data used in geological disaster study in Three Gorges, China. Elevation (left) and slope angle (right)

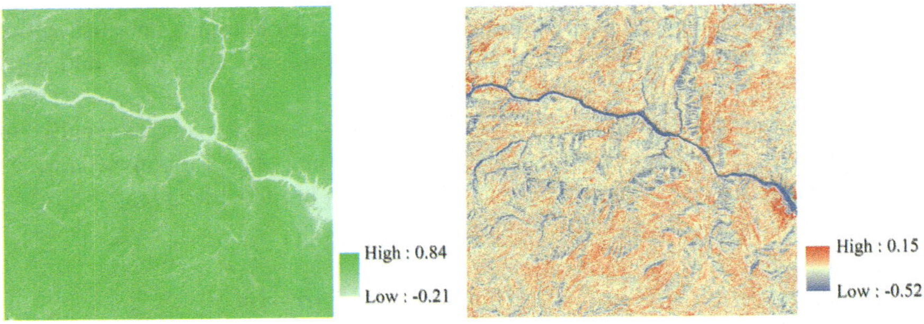


**Fig. 2.2** Vegetation and landcover features from optical remote sensing used in geological disaster study in Three Gorges, China. NDVI (left) and normalized difference build-up index (NDBI) (right)

debris flow occurrence (Baum & Godt, 2010; Terlien et al., 1995). Besides CNN for recognize the spatial pattern, RNN can be used to capture the temporal dependencies of these hydraulic features (Xu & Niu, 2018). It can process sequential data by maintaining a memory of previous inputs.

And landslides can be triggered by previous landslide events, so historical landslide data can be used as an important feature in RNN-based landslide prediction (Myronidis et al., 2016).

The underlying geology and soil can play a role in the geological disaster susceptibility, with certain rock and soil types being more prone to landslide, rockfall or collapse (Vandewater et al., 2005; Yalcin, 2007). Both geology and soil features can be obtained digital maps and remote sensing images, which can be then extracted by CNN.

(2) Ground subsidence, sinkhole, and ground fissure

Deep learning techniques can be used to recognize key features or patterns of these geological disasters by analyzing various types of data, including satellite imagery, LiDAR, InSAR and other geospatial data. Variation of land surface elevation is the key feature of these geological disasters which can be obtained from remote sensing data such as LiDAR or InSAR (Atzori et al., 2015; Wu et al., 2016). These data can provide high-resolution measurements of the land surface, so that CNN can be trained to identify patterns of elevation change over time, which can be indicative of areas where land subsidence or sinkholes are likely to occur.

Groundwater levels can be a key factor in the development of these geological disasters (Xu et al., 2012). When the groundwater is depleted, the soil particles lose their buoyancy and become compacted, leading to subsidence. Similarly, when the groundwater is overabundant, the soil particles can be eroded, leading to sinkholes. Groundwater level changes can also cause fissures to form in the ground due to the expansion and contraction of the soil (Da-yu & Li, 2000). RNN can be used to analyze time-series data groundwater levels to predict the likelihood of disaster occurring.

Deep learning can also analyze land use patterns, such as areas where land has been heavily developed or where there is significant agricultural activity, which can contribute to the instability of the ground and increase the likelihood of geological disasters (Amin & Bankher, 1997). In addition, geological features such as fault lines or underground caves, as well as soil properties such as water content, permeability, and porosity can also extracted by neural network to indicate areas where land subsidence or sinkholes may occur (Budhu & Adiyaman, 2013; Lu et al., 2020).

### *2.1.3 Detectability of Geological Disasters in Historical Image Change Analysis Based on Deep Learning*

Changes in time-series historical images can be used to identify and monitor geological disasters. Deep learning algorithms can be used to automatically extract features from image pairs taken at different times and detect changes between them. The image data sources to analysis the historical changes mainly include optical remote sensing and InSAR.

(1) Optical remote sensing

Current, a variety of optical remote sensing platforms that can be used for historical image change analysis, including Landsat series, Sentinel series, and MODIS. These platforms can provide a wealth of information about the earth surface, including changes in vegetation cover, land use and land cover changes. For example, changes in vegetation cover could indicate landslide and erosion (Alcántara-Ayala et al., 2006). And changes in the color or texture of the land surface could indicate landslide, subsidence, or ground fissure (Xun et al., 2022; Zhao et al., 2021). In addition, optical remote sensing data can also provide information about the physical properties, such as the reflectance, emissivity, and thermal properties. These properties can be used to analyze changes in soil moisture, land surface temperature, and other environmental factors that may be associated with geological disasters (Loche et al., 2022; Ray & Jacobs, 2007).

Deep learning algorithms can well identify and these changes obtained from optical remote sensing image data. CNNs can be used to analyze changes in geological structure, terrain, vegetation cover, land use (Shi et al., 2020), while RNN can be used to analyze time-series data, such as changes in precipitation or soil water content (Xu & Niu, 2018). These various architectures provide deep learning models the ability to learn complex patterns and relationships in multi-channels and large-size image data and make them well-suited for analyzing large volumes of optical remote sensing data to detect and analyze changes to identify geological disasters.

However, the detectability of geological disaster in historical optical remote sensing images can be influenced by several factors that related to inherent shortage of optical remote sensing. The spatial resolution determines the smallest detectable feature in the image and the temporal resolution determines the frequency at which images are acquired. Higher resolution imagery can reveal finer details and make it easier to identify smaller-scale geological features. And frequent image acquisition can help capture short-term changes in geological features, making it easier to detect hazards that are developing rapidly. So far most optical satellites cannot achieve the high spatial and temporal resolution simultaneously. In addition, cloud cover, haze, and other atmospheric conditions can obstruct or distort the view of the ground surface, making the available optical images data significantly decrease (Zhu et al., 2016). Vegetation cover can change seasonally throughout the year can make it challenging to monitoring the changes of geological features in some months (Yan et al., 2023). The shape and relief of the terrain can generate shadow that influences the appearance of geological features in the image, making it difficult to detect them in some areas (Amatya et al., 2019). The performance of deep learning can be restricted by abovementioned influences.

(2) InSAR

InSAR can provide information on the magnitude, location, and timing of the surface deformation. InSAR data is typically presented as an interferogram, which is a two-dimensional image that shows the difference in the distance between the satellite and the earth surface over time. The interferogram is created by combining two or more SAR images acquired at different times. The resulting image shows areas of ground deformation as fringes of different colors. By analyzing the interferogram, it is possible to detect ground deformation caused by geological disasters (Yang et al., 2020).

Deep learning can be used to analyze InSAR data by using different architectures for specific tasks. CNN can be trained to identify patterns of ground deformation and displacement in InSAR data (Anantrasirichai et al., 2019). RNNs can be used to analyze time series data of ground deformation and displacement, which can provide insights into the dynamics of geological processes (Kulshrestha et al., 2022). Autoencoders can also be used to extract features from InSAR data that can be used for ground deformation detection (Shakeel et al., 2022). The use of deep learning techniques can enable automatic and efficient processing of large amounts of InSAR data for change analysis of geological disasters.

The detectability of geological disasters by InSAR image change analysis depends on several factors, including the type of geological disaster, the spatiotemporal resolution of the InSAR data, and the accuracy of the deep learning algorithm. Different types of geological hazards exhibit different patterns of ground deformation in InSAR data. Landslides typically exhibit a characteristic pattern of downslope movement, while ground subsidence typically exhibits a pattern of sinking. The pattern can also be affected by other geological and environmental factors (Zhu et al., 2022). Therefore, the detectability depends on if the specific patterns and characteristics that they exhibit in the interferograms can be accurately extracted by applied deep learning model. The spatial and temporal resolution of InSAR data can also affect the detectability of geological disasters. Higher resolution data can capture smaller scale deformations, while data acquired at shorter intervals can capture faster deformation processes (Liu et al., 2018). The optimal resolution and interval of InSAR data depend on the specific geological disaster being monitored.

## 2.2 Methods of Intelligent Interpretation of Geological Disasters

The new generation of intelligent interpretation of geological disasters are mainly based on deep learning models constructed with artificial neural networks with many layers and various structures for specific functions. They are able to automatically learn complex representations of data by iteratively refining their internal representations through training on large datasets. The popular deep learning models in researches of geological disasters can be categorized to convolutional neural networks (CNN), deep generative models, recurrent neural network (RNN) and graph neural networks (GNN) (Ma & Mei, 2021), which will be detailly introduced in following sections.

### 2.2.1 *Convolutional Neural Networks*

Convolutional neural network (CNN) is a type of deep neural network that is commonly used for object detection, classification, and segmentation. The main characteristic of a CNN (Fig. 2.3) is the use of convolutional layers, which apply filters to the input image to extract features. These filters, also known as kernels or weights, are learned during the training process, allowing the network to identify complex patterns in the data. In addition to convolutional layers, a CNN also typically includes pooling layers, which downsample the feature maps generated by the convolutional layers, reducing the dimensionality of the data and increasing the network's efficiency. The output of the convolutional and pooling layers is then fed into one or more fully connected layers, which perform the final classification or regression task (O'Shea & Nash, 2015). The basic CNN architecture has been modified and improved to a series of popular architectures like AlexNet (Krizhevsky et al., 2012), VGGNet (Simonyan & Zisserman, 2014), GoogleNet (Szegedy et al., 2015), ResNet (He et al., 2016), etc. to achieve better performance.

U-Net is a CNN-based architecture designed for image segmentation tasks (Ronneberger et al., 2015). The U-Net architecture (Fig. 2.4) consists of an encoding path and a decoding path. The encoding path is similar to the convolutional layers in a

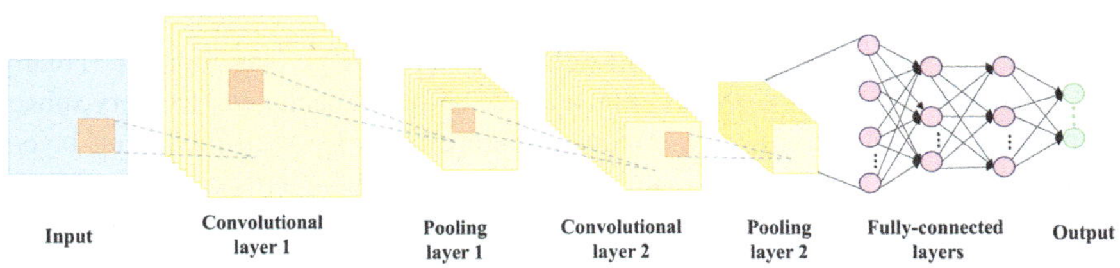


**Fig. 2.3** Structure of traditional convolutional neural network

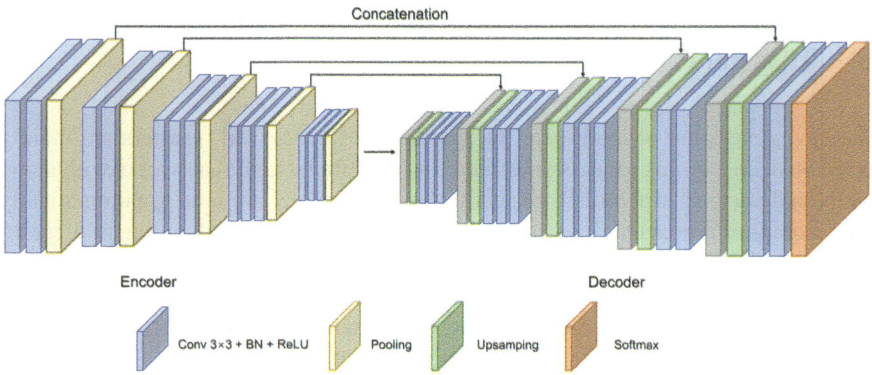


**Fig. 2.4** Structure of U-Net

typical CNN, where the input image is downsampled while features are extracted. The decoding path, on the other hand, consists of upsampling and concatenation layers, which increase the resolution of the feature map and combine it with the corresponding feature map from the encoding path. This allows the network to preserve high-resolution features while also capturing low-level features that are important for image segmentation. U-Net and its improved versions are widely used in the detection and extraction of various geological disasters including landslide (Chen et al., 2023; Liu et al., 2020; Meena et al., 2022; Soares et al., 2020), debris flow (Bai et al., 2021; Yokoya et al., 2020), sinkhole (Alshawi et al., 2023; Rafique et al., 2022), ground subsidence (Wang et al., 2021), collapse (Pan et al., 2022) and ground fissure (Xu et al., 2022a).

SegNet is another architecture designed for semantic segmentation, which involves assigning a class label to each pixel in an image (Badrinarayanan et al., 2017). It is also composed of an encoder-decoder structure, where the encoder consists of a series of convolutional and pooling layers, and the decoder consists of a series of deconvolutional layers (Fig. 2.5). One of the key features of SegNet is the use of pooling indices, which are the locations of the max values in the pooling layers and are saved during the encoding process and used during the decoding process to upsample the feature maps, effectively recovering the original resolution of the input image. There are also some researches using SegNet in geological disasters like detection and mapping of landslide (Antara et al., 2019; Yu et al., 2021), but much less than the U-Net based architectures.

DenseNet is an improved CNN algorithm based on image classification (Huang et al., 2017). The DenseNet connections feed the output of each layer to every subsequent layer, so that each layer can receive feature maps from all preceding layers and passes its own feature maps to all subsequent layers, which improves the flow of information and gradients throughout the network, allowing for better feature propagation and gradient updates. DenseNet-based landslide detection model has been proved to have better performance than traditional CNN-based model (Liu & Chen, 2021).

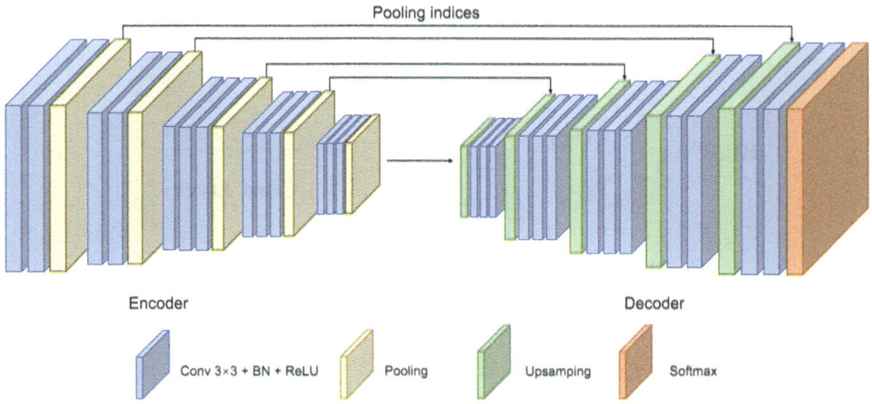


**Fig. 2.5** Structure of SegNet

R-CNN (Girshick et al., 2014) is a CNN-based object detection model. It includes two parts of region proposal and classification (Fig. 2.6). In the region proposal stage, the model generates region proposals using selective search, which is a segmentation-based object proposal algorithm. These proposals are then fed into a CNN to generate a fixed-length feature vector for each region proposal. In the classification stage, the feature vectors are used to classify each region proposal into one of the predefined classes. This is done using a support vector machine (SVM) classifier, which is trained to distinguish between the different classes based on the features extracted by the CNN. The slow processing time due to the need to run the CNN on each region proposal separately, which then led to the development of faster object detection models such as Fast R-CNN (Girshick, 2015), Faster R-CNN (Ren et al., 2015). And other improvements like Mask R-CNN (He et al., 2017), on the other hand, are designed to provide more accurate results. The R-CNN series are popular in detection tasks of geological disasters, like landslide (Fu et al., 2022; Liu et al., 2022; Ullo et al., 2021; Yang & Mao, 2022), rockfall (Chung & Yang, 2021) and ground subsidence (Wu et al., 2023).

YOLO (You Only Look Once) is another popular object detection system that can detect multiple objects within an image and predict their class labels and bounding boxes (Redmon et al., 2016). It has several improvements in terms of accuracy and speed and it is at the 8th version (YOLO v8) so far (Terven & Cordova-Esparza, 2023). Unlike other object detection systems that apply a classifier to various regions of an image, YOLO divides the image into a grid of cells and applies a CNN to each cell to predict bounding boxes, confidence scores, and class probabilities simultaneously. YOLO series models have also been widely used in the detections of landslide (Cheng et al., 2021; Guo et al., 2022; Li & Li, 2022), sinkhole (Kulkarni et al., 2023) and ground subsidence (Yu et al., 2022).

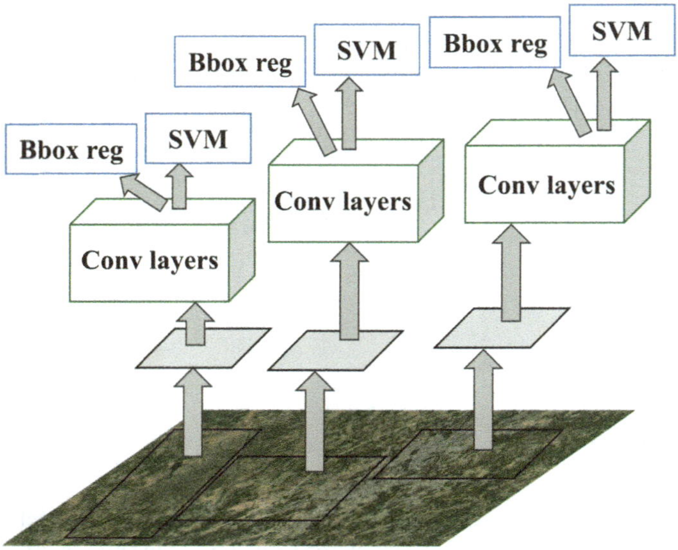


**Fig. 2.6** Structure of R-CNN

### *2.2.2 Deep Generative Models*

Deep generative model (DGM) is a type of artificial neural network with many hidden layers that is designed to learn the underlying structure of high-dimensional datasets, and generate new samples that indicate similar properties to the original data. It aims to capture the distribution of the data and generate new samples from that distribution. By using the multiple layers of non-linear transformations, DGM can learn increasingly complex features of the data as the depth of the network increases (Ruthotto & Haber, 2021).

Autoencoder is an unsupervised learning algorithm that can learn efficient data representations by reconstructing input data from a compressed latent space (Rumelhart et al., 1985). It consists of an encoder network that maps input data to a lower-dimensional latent space, and a decoder network that maps the latent space back to the input space (Fig. 2.7). An autoencoder can be trained to encode the features of a region such as topography, land use, and geological characteristics into a low-dimensional representation that captures the regional susceptibility to landslide (Huang et al., 2020; Nam & Wang, 2020).

Deep belief network (DBN) (Fig. 2.8) is another type of DGM that consists of multiple layers of restricted Boltzmann machines (RBMs) (Hinton et al., 2006). RBMs are unsupervised learning models that can learn a compressed representation of the input data. The output of one RBM is used as the input of the next RBM, allowing the DBN to learn increasingly complicated representations. DBN has been used in landslide detection (Ye et al., 2019), displacement prediction (Li et al., 2020a) and susceptibility mapping (Ghasemian et al., 2022; Xiong et al., 2021).

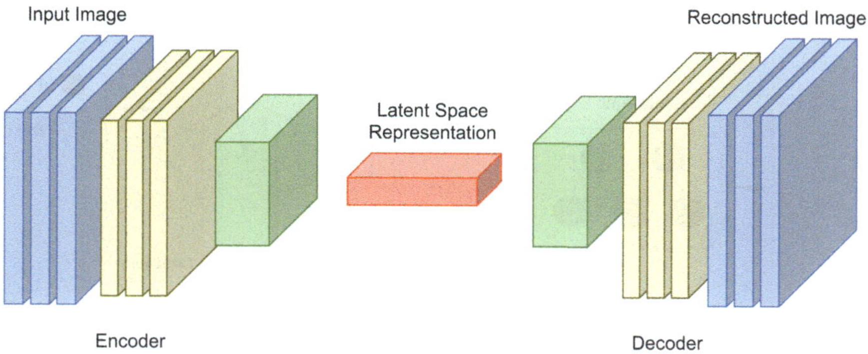


**Fig. 2.7** Structure of autoencoder

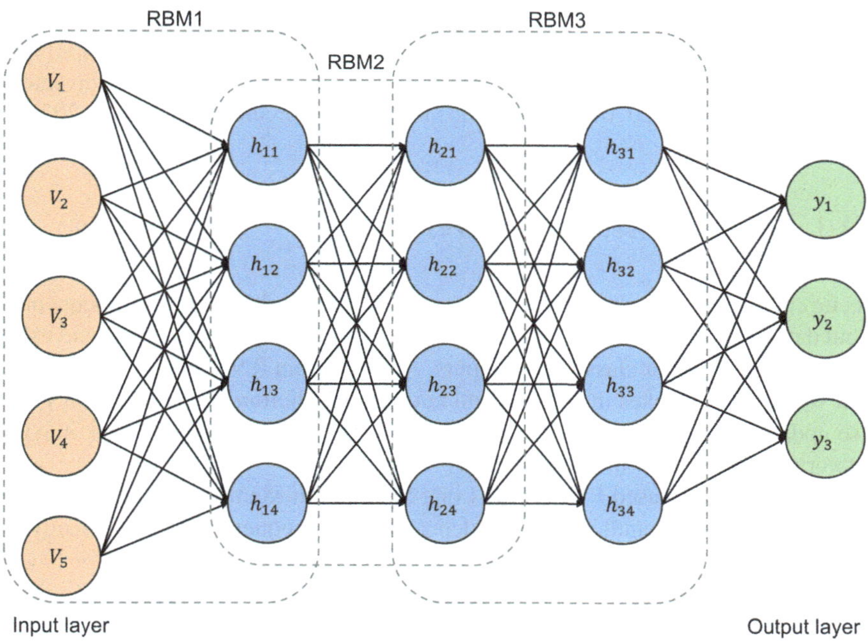


**Fig. 2.8** Structure of DBN

Generative adversarial network (GAN) is the recently proposed DGM and has been successfully used in various applications (Goodfellow et al., 2020). It consists of a generator and a discriminator (Fig. 2.9). The generator network is trained to create new data samples that resemble the training data, while the discriminator network is trained to distinguish between the actual and generated data. The two networks are trained in a competing fashion, where the generator tries to generate realistic samples that can fool the discriminator, and the discriminator tries to correctly identify which

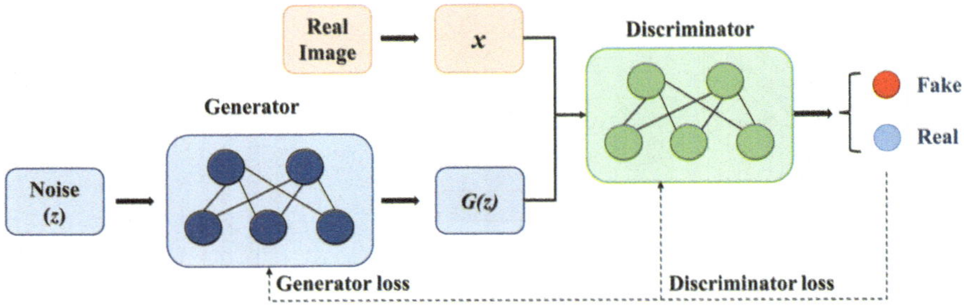


**Fig. 2.9** Structure of GAN

samples are real and which are generated. Through this competition, the generator network gradually improves its ability to generate realistic data samples that can fool the discriminator network. In researches of geological disasters, GAN can help generate additional data and solve the data imbalance issue for landslide susceptibility assessment (Al-Najjar & Pradhan, 2021; Al-Najjar et al., 2021), landslide inventory mapping (Fang et al., 2020) and landslide displacement prediction (Xu et al., 2022b).

## 2.2.3 *Recurrent Neural Networks*

Recurrent neural network (RNN) is a type of neural network working with sequential input data (Fig. 2.10). It can maintain a state or memory of previous inputs and use it to make predictions or classify new inputs. At its core, an RNN consists of a hidden state or cell, which takes in an input and outputs a prediction or classification while also updating its internal state. The output of each time step is then fed back into the network as an input for the next time step.

The most widely used RNN is the Long Short-Term Memory (LSTM) network, which is designed to handle long-term dependencies in sequential data by selectively

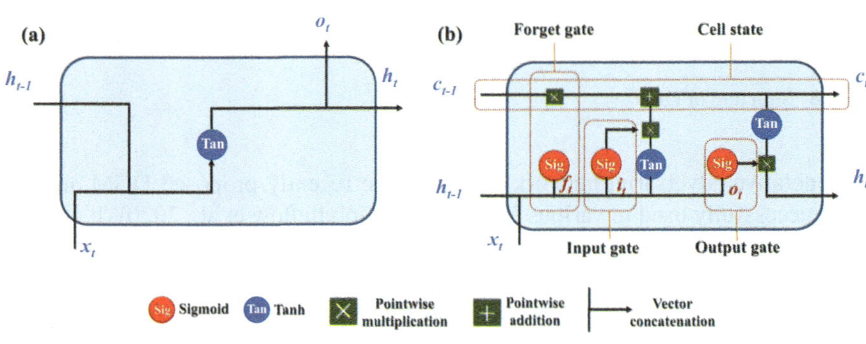


**Fig. 2.10** Structure of RNN (**a**) and LSTM (**b**)

remembering or forgetting previous inputs and thereby solve the issue of vanishing gradient in traditional RNN (Hochreiter & Schmidhuber, 1997). LSTM (Fig. 2.10) address this issue by introducing a memory cell and three gating mechanisms: the input gate, the output gate, and the forget gate. The memory cell allows the network to store information over long periods of time, while the gating mechanisms regulate how much information is stored and when it is forgotten. When a new input data is fed into the LSTM network, the input gate decides how much of the input should be added to the memory cell. The forget gate determines how much information should be removed from the memory cell. The output gate controls how much of the memory cell should be output to the next layer or as the final prediction. In researches of geological disaster, LSTM has been used in many researches, including displacement prediction of landslide (Li et al., 2021; Wang et al., 2022; Yang et al., 2019), earthquake prediction (Hasan Al Banna et al., 2021; Vardaan et al., 2019) and vulnerability assessment (Jena et al., 2021), and prediction of ground subsidence (Kumar et al., 2022; Liu et al., 2021).

### 2.2.4 Graph Neural Networks

Graph Neural Network (GNN) is a type of neural network that can operate on graph data structures rather than Euclidean data, where nodes and edges represent entities and relationships between them (Scarselli et al., 2008). GNN typically involve a message-passing scheme, where messages are exchanged between nodes and edges in a graph (Fig. 2.11). Each node updates its own representation based on its own features and the features of its neighboring nodes, which are aggregated to produce a new representation. This process is repeated for several iterations, allowing each node to integrate information from its neighbors and refine its representation.

Graph Convolutional Network (GCN) is the most popular type of GNN that have been used in studies of geological disasters, which extends convolutional neural

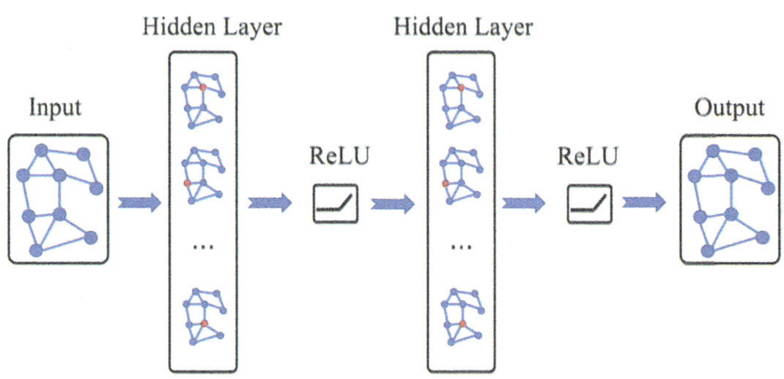


**Fig. 2.11** Structure of graph neural network

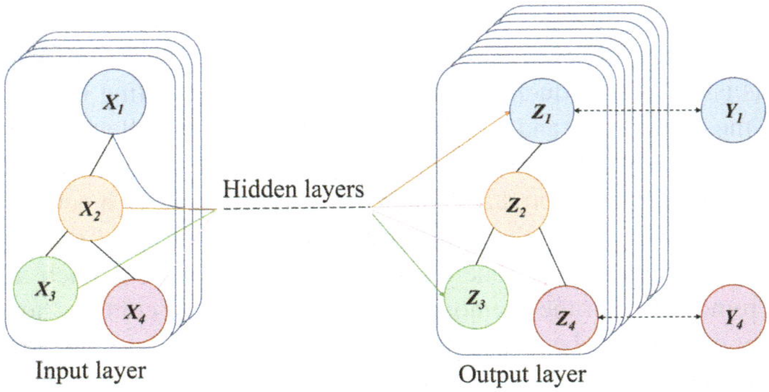


**Fig. 2.12** Graph convolutional network, modified from Kipf and Welling (2016)

networks to operate on graph-structured data. The GCN algorithm (Fig. 2.12) operates on a graph G = (V, E), where V is the set of nodes and E is the set of edges. Each node in the graph is associated with a feature vector, which encodes the properties of the node. The GCN learns a function that maps the feature vectors of the nodes to a new set of feature vectors by aggregating information from the node neighbors, which capture the properties of the nodes in a higher-dimensional space (Kipf & Welling, 2016). Studies on landslide deformation (Khalili et al., 2023; Ma et al., 2021) and susceptibility assessment (Wang et al., 2023) as well as earthquake detection and early warning (Bilal et al., 2022), which have considered GCN and achieved outstanding performance.

## References

Abdullah, A. Y., Masrur, A., Adnan, M. S., Baky, M. A., Hassan, Q. K., & Dewan, A. (2019). Spatio-temporal patterns of land use/land cover change in the heterogeneous coastal region of Bangladesh between 1990 and 2017. *Remote Sensing, 11*(7), 790.

Alcántara-Ayala, I., Esteban-Chávez, O., & Parrot, J. F. (2006). Landsliding related to land-cover change: A diachronic analysis of hillslope instability distribution in the Sierra Norte, Puebla, Mexico. *Catena, 65*(2), 152–165.

Al-Najjar, H. A., & Pradhan, B. (2021). Spatial landslide susceptibility assessment using machine learning techniques assisted by additional data created with generative adversarial networks. *Geoscience Frontiers, 12*(2), 625–637.

Al-Najjar, H. A., Pradhan, B., Sarkar, R., Beydoun, G., & Alamri, A. (2021). A new integrated approach for landslide data balancing and spatial prediction based on generative adversarial networks (GAN). *Remote Sensing, 13*(19), 4011.

Alshawi, R., Hoque, M. T., & Flanagin, M. C. (2023). A depth-wise separable U-Net architecture with multiscale filters to detect sinkholes. *Remote Sensing, 15*(5), 1384.

Amatya, P., Kirschbaum, D., & Stanley, T. (2019). Use of very high-resolution optical data for landslide mapping and susceptibility analysis along the Karnali highway, Nepal. *Remote Sensing, 11*(19), 2284.

Amin, A., & Bankher, K. (1997). Causes of land subsidence in the Kingdom of Saudi Arabia. *Natural Hazards, 16*, 57–63.

Anantrasirichai, N., Biggs, J., Albino, F., & Bull, D. (2019). The application of convolutional neural networks to detect slow, sustained deformation in InSAR time series.' *Geophysical Research Letters, 46*(21), 11850–11858.

Antara, I. M., Shimizu, N., Osawa, T., & Nuarsa, I. W. (2019). An application of SegNet for detecting landslide areas by using fully polarimetric SAR data. *Ecotrophic, 13*(2), 215–226.

Atzori, S., Baer, G., Antonioli, A., & Salvi, S. (2015). InSAR-based modeling and analysis of sinkholes along the Dead Sea coastline. *Geophysical Research Letters, 42*(20), 8383–8390.

Badrinarayanan, V., Kendall, A., & Cipolla, R. (2017). SegNet: A deep convolutional encoder-decoder architecture for image segmentation. *IEEE Transactions on Pattern Analysis and Machine Intelligence, 39*(12), 2481–2495.

Bai, T., Jiang, Z., & Tahmasebi, P. (2021). Debris flow prediction with machine learning: Smart management of urban systems and infrastructures. *Neural Computing and Applications, 33*(22), 15769–15779.

Baum, R. L., & Godt, J. W. (2010). Early warning of rainfall-induced shallow landslides and debris flows in the USA. *Landslides, 7*, 259–272.

Bilal, M. A., Ji, Y., Wang, Y., Akhter, M. P., & Yaqub, M. (2022). Early earthquake detection using batch normalization graph convolutional neural network (BNGCNN). *Applied Sciences, 12*(15), 7548.

Budhu, M., & Adiyaman, I. (2013). The influence of clay zones on land subsidence from groundwater pumping. *Groundwater, 51*(1), 51–57.

Chen, H., He, Y., Zhang, L., Yao, S., Yang, W., Fang, Y., Liu, Y., & Gao, B. (2023). A landslide extraction method of channel attention mechanism U-Net network based on Sentinel-2A remote sensing images. *International Journal of Digital Earth, 16*(1), 552–577.

Chen, Y., Jiang, H., Li, C., Jia, X., & Ghamisi, P. (2016). Deep feature extraction and classification of hyperspectral images based on convolutional neural networks. *IEEE Transactions on Geoscience and Remote Sensing, 54*(10), 6232–6251.

Cheng, L., Li, J., Duan, P., & Wang, M. (2021). A small attentional YOLO model for landslide detection from satellite remote sensing images. *Landslides, 18*(8), 2751–2765.

Chung, Y.-L., & Yang, J.-J. (2021). Application of a mask R-CNN-based deep learning model combined with the retinex image enhancement algorithm for detecting rockfall and potholes on hill roads. In *2021 IEEE 11th International Conference on Consumer Electronics (ICCE-Berlin)* (pp. 1–6).

Da-yu, G., & Li, Z. (2000). Ground fissure hazards in USA and China. *Acta Seismologica Sinica, 13*(4), 466–476.

Fang, B., Chen, G., Pan, L., Kou, R., & Wang, L. (2020). GAN-based Siamese framework for landslide inventory mapping using bi-temporal optical remote sensing images. *IEEE Geoscience and Remote Sensing Letters, 18*(3), 391–395.

Fu, R., He, J., Liu, G., Li, W., Mao, J., He, M., & Lin, Y. (2022). Fast seismic landslide detection based on improved mask R-CNN. *Remote Sensing, 14*(16), 3928.

Ghasemian, B., Shahabi, H., Shirzadi, A., Al-Ansari, N., Jaafari, A., Kress, V. R., Geertsema, M., Renoud, S., & Ahmad, A. (2022). A robust deep-learning model for landslide susceptibility mapping: A case study of Kurdistan Province, Iran. *Sensors, 22*(4), 1573.

Girshick, R. (2015). Fast R-CNN. In *Proceedings of the IEEE International Conference on Computer Vision* (pp. 1440–1448).

Girshick, R., Donahue, J., Darrell, T., & Malik, J. (2014). Rich feature hierarchies for accurate object detection and semantic segmentation. In *Proceedings of the IEEE Conference on Computer Vision and Pattern Recognition* (pp. 580–587).

Glade, T. (2003). Landslide occurrence as a response to land use change: A review of evidence from New Zealand. *CATENA, 51*(3–4), 297–314.

Goodfellow, I., Pouget-Abadie, J., Mirza, M., Xu, B., Warde-Farley, D., Ozair, S., Courville, A., & Bengio, Y. (2020). Generative adversarial networks. *Communications of the ACM, 63*(11), 139–144.

Guo, H., Yi, B., Yao, Q., Gao, P., Li, H., Sun, J., & Zhong, C. (2022). Identification of landslides in mountainous area with the combination of SBAS-InSAR and YOLO model. *Sensors, 22*(16), 6235.

Hasan Al Banna, M., Ghosh, T., Taher, K. A., Kaiser, M. S., & Mahmud, M. (2021). An earthquake prediction system for Bangladesh using deep long short-term memory architecture. In *Intelligent Systems: Proceedings of ICMIB 2020* (pp. 465–476).

He, K., Gkioxari, G., Dollár, P., & Girshick, R. (2017). Mask R-CNN. In *Proceedings of the IEEE International Conference on Computer Vision* (pp. 2961–2969).

He, K., Zhang, X., Ren, S., & Sun, J. (2016). Deep residual learning for image recognition. In *Proceedings of the IEEE Conference on Computer Vision and Pattern Recognition* (pp. 770–778).

Hinton, G. E., Osindero, S., & Teh, Y. W. (2006). A fast learning algorithm for deep belief nets. *Neural Computation, 18*(7), 1527–1554.

Hochreiter, S., & Schmidhuber, J. (1997). Long short-term memory. *Neural Computation, 9*(8), 1735–1780.

Huang, F., Zhang, J., Zhou, C., Wang, Y., Huang, J., & Zhu, L. (2020). A deep learning algorithm using a fully connected sparse autoencoder neural network for landslide susceptibility prediction. *Landslides, 17*, 217–229.

Huang, G., Liu, Z., Van Der Maaten, L., & Weinberger, K. Q. (2017). Densely connected convolutional networks. In *Proceedings of the IEEE Conference on Computer Vision and Pattern Recognition* (pp. 4700–4708).

Jena, R., Naik, S. P., Pradhan, B., Beydoun, G., Park, H. J., & Alamri, A. (2021). Earthquake vulnerability assessment for the Indian subcontinent using the Long Short-Term Memory model (LSTM). *International Journal of Disaster Risk Reduction, 66*, 102642.

Jogin, M., Madhulika, M. S., Divya, G. D., Meghana, R. K., & Apoorva, S. (2018). Feature extraction using convolution neural networks (CNN) and deep learning. In *2018 3rd IEEE International Conference on Recent Trends in Electronics, Information & Communication Technology (RTEICT)* (pp. 2319–2323).

Kakogeorgiou, I., & Karantzalos, K. (2021). Evaluating explainable artificial intelligence methods for multi-label deep learning classification tasks in remote sensing. *International Journal of Applied Earth Observation and Geoinformation, 103*, 102520.

Khalili, M. A., Guerriero, L., Pouralizadeh, M., Calcaterra, D., & Di Martire, D. (2023). Prediction of deformation caused by landslides based on graph convolution networks algorithm and DInSAR technique. *ISPRS Annals of the Photogrammetry, Remote Sensing and Spatial Information Sciences, 10*, 391–397.

Kipf, T. N., & Welling, M. (2016). Semi-supervised classification with graph convolutional networks. arXiv preprint, arXiv:1609.02907

Krizhevsky, A., Sutskever, I., & Hinton, G. E. (2012). ImageNet classification with deep convolutional neural networks. *Advances in Neural Information Processing Systems, 25*, 1–9.

Kulkarni, N. N., Raisi, K., Valente, N. A., Benoit, J., Yu, T., & Sabato, A. (2023). Deep learning augmented infrared thermography for unmanned aerial vehicles structural health monitoring of roadways. *Automation in Construction, 148*, 104784.

Kulshrestha, A., Chang, L., & Stein, A. (2022). Use of LSTM for sinkhole-related anomaly detection and classification of InSAR deformation time series. *IEEE Journal of Selected Topics in Applied Earth Observations and Remote Sensing, 15*, 4559–4570.

Kumar, S., Kumar, D., Donta, P. K., & Amgoth, T. (2022). Land subsidence prediction using recurrent neural networks. *Stochastic Environmental Research and Risk Assessment, 36*(2), 373–388.

Lee, S., & Talib, J. A. (2005). Probabilistic landslide susceptibility and factor effect analysis. *Environmental Geology, 47*, 982–990.

Li, B., & Li, J. (2022). Methods for landslide detection based on lightweight YOLOv4 convolutional neural network. *Earth Science Informatics, 15*(2), 765–775.

Li, H., Xu, Q., He, Y., Fan, X., & Li, S. (2020a). Modeling and predicting reservoir landslide displacement with deep belief network and EWMA control charts: A case study in Three Gorges Reservoir. *Landslides, 17*(3), 693–707.

Li, W., Fan, X., Huang, F., Chen, W., Hong, H., Huang, J., & Guo, Z. (2020b). Uncertainties analysis of collapse susceptibility prediction based on remote sensing and GIS: Influences of different data-based models and connections between collapses and environmental factors. *Remote Sensing, 24*, 4134.

Li, L. M., Zhang, M. Y., & Wen, Z. Z. (2021). Dynamic prediction of landslide displacement using singular spectrum analysis and stack long short-term memory network. *Journal of Mountain Science, 18*(10), 2597–2611.

Liang, H., Sun, X., Sun, Y., & Gao, Y. (2017). Text feature extraction based on deep learning: A review. *EURASIP Journal on Wireless Communications and Networking*, 1–12.

Liu, N., Dai, W., Santerre, R., Hu, J., Shi, Q., & Yang, C. (2018). High spatio-temporal resolution deformation time series with the fusion of InSAR and GNSS data using spatio-temporal random effect model. *IEEE Transactions on Geoscience and Remote Sensing, 57*, 364–380.

Liu, P., Wei, Y., Wang, Q., Chen, Y., & Xie, J. (2020). Research on post-earthquake landslide extraction algorithm based on improved U-Net model. *Remote Sensing, 12*(5), 894.

Liu, Q., Zhang, Y., Wei, J., Wu, H., & Deng, M. (2021). HLSTM: Heterogeneous long short-term memory network for large-scale InSAR ground subsidence prediction. *IEEE Journal of Selected Topics in Applied Earth Observations and Remote Sensing, 14*, 8679–8688.

Liu, T., & Chen, T. (2021). A comparation of CNN and DenseNet for landslide detection. In *2021 IEEE International Geoscience and Remote Sensing Symposium IGARSS* (pp. 8440–8443).

Liu, Y., Yao, X., Gu, Z., Zhou, Z., Liu, X., Chen, X., & Wei, S. (2022). Study of the automatic recognition of landslides by using InSAR images and the improved mask R-CNN model in the Eastern Tibet Plateau. *Remote Sensing, 14*(14), 3362.

Loche, M., Scaringi, G., Yunus, A. P., Catani, F., Tanyaş, H., Frodella, W., Fan, X., & Lombardo, L. (2022). Surface temperature controls the pattern of post-earthquake landslide activity. *Scientific Reports, 12*(1), 988.

Lu, Q., Liu, Y., Peng, J., Li, L., Fan, W., Liu, N., Sun, K., & Liu, R. (2020). Immersion test of loess in ground fissures in Shuanghuaishu, Shanxi Province, China. *Bulletin of Engineering Geology and the Environment, 79*, 2299–2312.

Lv, N., Chen, C., Qiu, T., & Sangaiah, A. K. (2018). Deep learning and superpixel feature extraction based on contractive autoencoder for change detection in SAR images. *IEEE Transactions on Industrial Informatics, 14*(12), 5530–5538.

Ma, A., Filippi, A. M., Wang, Z., Yin, Z., Huo, D., Li, X., & Güneralp, B. (2020). Fast sequential feature extraction for recurrent neural network-based hyperspectral image classification. *IEEE Transactions on Geoscience and Remote Sensing, 59*(7), 5920–5937.

Ma, Z., & Mei, G. (2021). Deep learning for geological hazards analysis: Data, models, applications, and opportunities. *Earth-Science Reviews, 223*, 103858.

Ma, Z., Mei, G., Prezioso, E., Zhang, Z., & Xu, N. (2021). A deep learning approach using graph convolutional networks for slope deformation prediction based on time-series displacement data. *Neural Computing and Applications, 33*(21), 14441–14457.

Mayer, R., & Jacobsen, H.-A. (2020). Scalable deep learning on distributed infrastructures: Challenges, techniques, and tools. *ACM Computing Surveys (CSUR), 53*(1), 1–37.

Meena, S. R., Soares, L. P., Grohmann, C. H., Van Westen, C., Bhuyan, K., Singh, R. P., Floris, M., & Catani, F. (2022). Landslide detection in the Himalayas using machine learning algorithms and U-Net. *Landslides, 19*(5), 1209–1229.

Myronidis, D., Papageorgiou, C., & Theophanous, S. (2016). Landslide susceptibility mapping based on landslide history and analytic hierarchy process (AHP). *Natural Hazards, 81*, 245–263.

Nam, K., & Wang, F. (2020). An extreme rainfall-induced landslide susceptibility assessment using autoencoder combined with random forest in Shimane Prefecture, Japan. *Geoenvironmental Disasters, 7*(1), 1–16.

O'Shea, K., & Nash, R. (2015). An introduction to convolutional neural networks. arXiv preprint, arXiv:1511.08458

Pan, H., Qin, S., Liu, G., Meng, F., Xiong, L., Yao, J., & Qiao, S. (2022). A collapse extraction method of remote sensing image using improved U-Net convolution network. *Journal of Research in Science and Engineering, 4*(10), 50–55.

Rafique, M. U., Zhu, J., & Jacobs, N. (2022). Automatic segmentation of sinkholes using a convolutional neural network. *Earth and Space Science, 9*(2), e2021EA002195.

Ray, R. L., & Jacobs, J. M. (2007). Relationships among remotely sensed soil moisture, precipitation and landslide events. *Natural Hazards, 43*, 211–222.

Redmon, J., Divvala, S., Girshick, R., & Farhadi, A. (2016). You only look once: Unified, real-time object detection. In *Proceedings of the IEEE Conference on Computer Vision and Pattern Recognition* (pp. 779–788).

Ren, S., He, K., Girshick, R., & Sun, J. (2015). Faster R-CNN: Towards real-time object detection with region proposal networks. *Advances in Neural Information Processing Systems,* 28.

Romero, A., Gatta, C., & Camps-Valls, G. (2015). Unsupervised deep feature extraction for remote sensing image classification. *IEEE Transactions on Geoscience and Remote Sensing, 54*(3), 1349–1362.

Ronneberger, O., Fischer, P., & Brox, T. (2015). U-Net: Convolutional networks for biomedical image segmentation. In *Medical Image Computing and Computer-Assisted Intervention—MICCAI 2015: 18th International Conference*, Munich, Germany, October 5–9, 2015, Proceedings, Part III 18 (pp. 234–241).

Rumelhart, D. E., Hinton, G. E., & Williams, R. J. (1985). *Learning internal representations by error propagation*. University of California, San Diego, Department of Cognitive Science.

Ruthotto, L., & Haber, E. (2021). An introduction to deep generative modeling. *GAMM-Mitteilungen, 44*(2), e202100008.

Scarselli, F., Gori, M., Tsoi, A. C., Hagenbuchner, M., & Monfardini, G. (2008). The graph neural network model. *IEEE Transactions on Neural Networks, 20*(1), 61–80.

Shakeel, A., Walters, R. J., Ebmeier, S. K., & Al Moubayed, N. (2022). ALADDIn: Autoencoder-LSTM-based anomaly detector of deformation in InSAR. *IEEE Transactions on Geoscience and Remote Sensing, 60*, 1–12.

Shi, W., Zhang, M., Ke, H., Fang, X., Zhan, Z., & Chen, S. (2020). Landslide recognition by deep convolutional neural network and change detection. *IEEE Transactions on Geoscience and Remote Sensing, 59*(6), 4654–4672.

Shirzadi, A., Saro, L., Hyun Joo, O., & Chapi, K. (2012). A GIS-based logistic regression model in rock-fall susceptibility mapping along a mountainous road: Salavat Abad case study, Kurdistan, Iran. *Natural Hazards, 64*, 1639–1656.

Simonyan, K., & Zisserman, A. (2014). Very deep convolutional networks for large-scale image recognition. arXiv preprint, arXiv:1409.1556

Soares, L. P., Dias, H. C., & Grohmann, C. H. (2020). Landslide segmentation with U-Net: Evaluating different sampling methods and patch sizes. arXiv preprint, arXiv:2007.06672

Szegedy, C., Liu, W., Jia, Y., Sermanet, P., Reed, S., Anguelov, D., Erhan, D., Vanhoucke, V., & Rabinovich, A. (2015). Going deeper with convolutions. In *Proceedings of the IEEE Conference on Computer Vision and Pattern Recognition* (pp. 1–9).

Terlien, M. T., Van Westen, C. J., & van Asch, T. W. (1995). Deterministic modelling in GIS-based landslide hazard assessment. In *Geographical information systems in assessing natural hazards* (pp. 57–77).

Terven, J., & Cordova-Esparza, D. (2023). A comprehensive review of YOLO: From YOLOv1 to YOLOv8 and beyond. arXiv preprint, arXiv:2304.00501

Ullo, S. L., Mohan, A., Sebastianelli, A., Ahamed, S. E., Kumar, B., Dwivedi, R., & Sinha, G. R. (2021). A new mask R-CNN-based method for improved landslide detection. *IEEE Journal of Selected Topics in Applied Earth Observations and Remote Sensing, 14*, 3799–3810.

Vandewater, C. J., Dunne, W. M., Mauldon, M., Drumm, E. C., & Bateman, V. (2005). Classifying and assessing the geologic contribution to rockfall hazard. *Environmental & Engineering Geoscience, 11*(2), 141–154.

Vardaan, K., Bhandarkar, T., Satish, N., Sridhar, S., Sivakumar, R., & Ghosh, S. (2019). Earthquake trend prediction using long short-term memory RNN. *International Journal of Electrical and Computer Engineering, 9*(2), 1304–1312.

Wang, X., Du, A., Hu, F., Liu, Z., Zhang, X., Wang, L., & Guo, H. (2023). Landslide susceptibility evaluation based on active deformation and graph convolutional network algorithm. *Frontiers in Earth Science, 11*, 1132722.

Wang, Y., Tang, H., Huang, J., Wen, T., Ma, J., & Zhang, J. (2022). A comparative study of different machine learning methods for reservoir landslide displacement prediction. *Engineering Geology, 298*, 106544.

Wang, Z., Li, L., Yu, Y., Wang, J., Li, Z., & Liu, W. (2021). A novel phase unwrapping method used for monitoring the land subsidence in coal mining area based on U-Net convolutional neural network. *Frontiers in Earth Science, 9*, 761653.

Wu, Q., Deng, C., & Chen, Z. (2016). Automated delineation of karst sinkholes from LiDAR-derived digital elevation models. *Geomorphology, 266*, 1–10.

Wu, Z., Ma, P., Zheng, Y., Gu, F., Liu, L., & Lin, H. (2023). Automatic detection and classification of land subsidence in deltaic metropolitan areas using distributed scatterer InSAR and oriented R-CNN. *Remote Sensing of Environment, 290*, 113545.

Xie, F., Wen, H., Wu, J., Chen, S., Hou, W., & Jiang, Y. (2019). Convolution based feature extraction for edge computing access authentication. *IEEE Transactions on Network Science and Engineering, 7*(4), 2336–2346.

Xiong, Y., Zhou, Y., Wang, F., Wang, S., Wang, J., Ji, J., & Wang, Z. (2021). Landslide susceptibility mapping using ant colony optimization strategy and deep belief network in Jiuzhaigou Region. *IEEE Journal of Selected Topics in Applied Earth Observations and Remote Sensing, 14*, 11042–11057.

Xu, J. J., Zhang, H., Tang, C. S., Cheng, Q., Liu, B., & Bin, S. (2022a). Automatic soil desiccation crack recognition using deep learning. *Geotechnique, 72*(4), 337–349.

Xu, M., Chen, J., Yang, H., & Xiao, T. (2022b). Combined with decomposition algorithm and generative adversarial networks on landslide displacement prediction. In *IEEE 2022 14th International Conference on Advanced Computational Intelligence (ICACI)* (pp. 42–48).

Xu, S., & Niu, R. (2018). Displacement prediction of Baijiabao landslide based on empirical mode decomposition and long short-term memory neural network in Three Gorges area, China. *Computers & Geosciences, 111*, 87–96.

Xu, Y.-S., Ma, L., Du, Y.-J., & Shen, S.-L. (2012). Analysis of urbanisation-induced land subsidence in Shanghai. *Natural Hazards, 63*, 1255–1267.

Xun, Z., Zhao, C., Kang, Y., Liu, X., Liu, Y., & Du, C. (2022). Automatic extraction of potential landslides by integrating an optical remote sensing image with an InSAR-derived deformation map. *Remote Sensing, 14*(11), 2669.

Yalcin, A. (2007). The effects of clay on landslides: A case study. *Applied Clay Science, 38*(1–2), 77–85.

Yan, L., Gong, Q., Wang, F., Chen, L., Li, D., & Yin, K. (2023). Integrated methodology for potential landslide identification in highly vegetation-covered areas. *Remote Sensing, 15*(6), 1518.

Yang, B., Yin, K., Lacasse, S., & Liu, Z. (2019). Time series analysis and long short-term memory neural network to predict landslide displacement. *Landslides, 16*, 677–694.

Yang, D., & Mao, Y. (2022). Remote sensing landslide target detection method based on improved faster R-CNN. *Journal of Applied Remote Sensing, 16*(4), 044521.

Yang, Z., Li, Z., Zhu, J., Wang, Y., & Wu, L. (2020). Use of SAR/InSAR in mining deformation monitoring, parameter inversion, and forward predictions: A review. *IEEE Geoscience and Remote Sensing Magazine, 8*(1), 71–90.

Ye, C., Li, Y., Cui, P., Liang, L., Pirasteh, S., Marcato, J., Goncalves, W. N., & Li, J. (2019). Landslide detection of hyperspectral remote sensing data based on deep learning with constrains.

*IEEE Journal of Selected Topics in Applied Earth Observations and Remote Sensing, 12*(12), 5047–5060.

Ying, X. (2019). An overview of overfitting and its solutions. *Journal of Physics: Conference Series*, 022022.

Yokoya, N., Yamanoi, K., He, W., Baier, G., Adriano, B., Miura, H., & Oishi, S. (2020). Breaking limits of remote sensing by deep learning from simulated data for flood and debris-flow mapping. *IEEE Transactions on Geoscience and Remote Sensing, 60*, 1–15.

Yu, B., Chen, F., Xu, C., Wang, L., & Wang, N. (2021). Matrix SegNet: A practical deep learning framework for landslide mapping from images of different areas with different spatial resolutions. *Remote Sensing, 13*(16), 3158.

Yu, Y., Wang, Z., Li, Z., Ye, K., Li, H., & Wang, Z. (2022). A lightweight anchor-free subsidence basin detection model with adaptive sample assignment in interferometric synthetic aperture radar interferogram. *Frontiers in Ecology and Evolution, 10*, 158.

Zhao, Y., Sun, B., Liu, S., Zhang, C., He, X., Xu, D., & Tang, W. (2021). Identification of mining induced ground fissures using UAV and infrared thermal imager: Temperature variation and fissure evolution. *ISPRS Journal of Photogrammetry and Remote Sensing, 180*, 45–64.

Zhu, K., Zhang, X., Sun, Q., Wang, H., & Hu, J. (2022). Characterizing spatiotemporal patterns of land deformation in the Santa Ana Basin, Los Angeles, from InSAR time series and independent component analysis. *Remote Sensing, 14*(11), 2624.

Zhu, X., Helmer, E. H., Gao, F., Liu, D., Chen, J., & Lefsky, M. A. (2016). A flexible spatiotemporal method for fusing satellite images with different resolutions. *Remote Sensing of Environment, 172*, 165–177.

# Chapter 3
# Intelligent Analysis of Multi-source Long-Term Landslide Ground Monitoring Data

**Abstract** Landslide displacement is an important indicator for early warning. It is of great importance to predict the landslide displacement for its prevention and mitigation. As displacement can be treated as a time series data, machine learning methods that can process time series data can be used for its prediction with related environmental variables. This chapter used RF and LSTM to predict landslide long-term time series rainfall and variation of reservoir water level were included as the environmental covariates of measured displacement. STL and Exponential smoothing methods were used to process the time series data and improve the model performance. The results indicate both RF and LSTM can successfully predict the landslide displacement.

## 3.1 Introduction

### *3.1.1 Background and Significance*

China is of large mountainous areas with complicated topographical structures and well-developed tectonics. Geological disasters are widely distributed in this country, which not only cause significant damage to the ecological environment, but also seriously threaten the life and property safety of local residents (Xie et al., 2021). Landslides account for the largest proportion of geological disasters, and landslide displacement prediction is one of the most important tasks for landslide control and prevention. Generally speaking, enhancing the research on landslide displacement can clearly explore the mechanism and rule of the landslide formation, ensure the stability of water resources, land use and ecosystem, which thereby promote infrastructure construction and planning. From a micro aspect, timely and accurate prediction of landslide displacement can help people take necessary preventive measures to ensure their safety of lives and properties.

In recent years, a variety of machine learning algorithms have been widely used in landslide spatial assessment and prediction, such as artificial neural network (ANN), decision tree (DT), support vector machine (SVM), random forest (RF), etc. (Zhou et al., 2022). However, the landslide displacement prediction in regional scale was

 
W. Chen et al., *Intelligent Interpretation for Geological Disasters*,
https://doi.org/10.1007/978-981-99-5822-1_3

rarely mentioned. The main application direction of machine learning algorithms is prediction and classification problems, which is suitable for displacement prediction of regional landslide. In addition, the quantity and quality of training samples determine the prediction accuracy of the early warning model to a certain extent. With the accumulation of big data including the historical landslide records, geological environment survey data, and landslide inducing factors over the years, the foundation has been well-laid for the development of landslide displacement prediction models based on data-driven machine learning methods. Considering that the prediction of landslide displacement is still challenging due to its complexity and dynamics, it is interdisciplinary research that requires the integration of geology, geomorphology, and engineering.

### 3.1.2 Research Overview

At present, researchers keep using and improving the machine learning methods for landslide displacement prediction. Zhou et al. (2015) proposed a wavelet decomposition-extreme learning machine (WA-ELM) landslide displacement prediction model based on chaotic time series to deal with the issue of chaotic characteristics of landslide displacement sequences and overcome the shortcomings of traditional time series prediction models. By analyzing the chaotic characteristics of the landslide displacement sequence, the model used wavelet analysis to decompose the displacement sequence into components with different frequency characteristics. The phase space reconstruction of each feature component was carried out and the extreme learning machine was used for prediction. Finally, the predicted values of each feature component were superimposed to obtain the prediction of the original displacement sequence. Compared with the wavelet analysis-support vector machine (WA-SVM) and the single ELM model, the prediction results indicated better universality and stability. Yan et al. (2021) proposed a method of combining time series and gated recurrent neural network (GRNN) to predict landslide displacement. The moving average method was used to decompose the total displacement curve of the landslide into trend item displacement and periodic item displacement. The gray Verhulst model described the trend item change. Influences of rainfall and reservoir water level on landslide displacement were considered for the prediction of periodic item change. The comparison of outputs with measured data indicated that the modified model had a better performance. Liang et al. (2022) used long short-term memory networks (LSTM) and recurrent neural network (RNN) to predict the landslide displacement based on the monitoring data of the Bazimen landslide for 10 years. In the study, the total displacement was decomposed into trend items and periodic items by using the moving average method, and the trend items were predicted by segmental fitting with cubic polynomial functions. The relationship between periodic items and eigenfactors were constructed for the prediction of abovementioned neural networks. The eigenfactors of the periodic item were initially extracted according to the factors affecting the displacement. Then the irrelevant factors were removed by using the Pearson correlation analysis. The predicted value of the total displacement was obtained by integrating the predicted trend items and periodic items. The error analysis between the predicted value and the real value was performed and indicated that LSTM is more suitable for long-term series prediction. Wang et al. (2022) proposed a dynamic prediction method of landslide displacement based on time series and convolutional long-short-term memory (CNN-LSTM) hybrid neural network. Compared with traditional BP, GRNN, LSTM and other methods, accuracy of the new network was significant improvement. However, the prediction model used was designed for the whole evolution process of landslide, and did not consider the different evolution stages of the landslide mass. Selecting corresponding models for different stages of landslide would be the focus of future researches.

Overall, including multi-source data allows the models to receive more information related to landslide displacement and make it perform better. Additionally, since landslide displacement prediction is a typical time series problem, there is a dynamic regularity between the data, and the displacement state at the previous moment can affect the following displacement state. Therefore, the neural networks that can process time series data will perform better in this task.

### *3.1.3 Research Object and Contents*

According to the reviewed literatures, the main issues related to landslide prediction is the lower prediction accuracy caused by single input feature gained from in-situ monitoring. And the traditional static neural network model cannot indicate the dynamic regularity of time series data. To tackle these issues, the main contents of this paper are as follows:

(1) The long-term time series rainfall and variation of reservoir water level are included to constructed the sample set of regional landslides.
(2) Static tree-based RF model and LSTM model are used in this study. As a special recurrent neural network, LSTM can selectively retain the information of the previous several moments and meet the requirements of dynamic prediction of time series (Gers et al., 2002). Therefore, it is used to predict landslide displacement in this study. LSTM network will be trained on the abovementioned sample set.

The content structure of this chapter is arranged as follows: Sect. 3.2 introduces technologies used in this study, including RF and LSTM. Section 3.3 introduces the acquisition of datasets and the process of model construction. Section 3.4 is the results and analysis of the experiment. Section 3.5 summarizes the research and proposes improvement suggestions for the problems of current research method.

## 3.2 Related Principles and Techniques

### 3.2.1 Random Forest

RF is a tree-based bagging machine learning algorithm that can be used for both classification and regression (Breiman, 2001). It works by constructing a number of decision trees and making them work together to reduce the error rate of the single decision tree. In the process of constructing the single decision tree, the algorithm will randomly select parts of features for the model training to avoid the overfitting, thereby improving the generalization ability of the model. RF model includes multiple decision trees and votes of these decision trees will be integrated to obtain the final classification or prediction result. The decision process is shown in Fig. 3.1. This approach tends to improve prediction accuracy and is also effective for handling outliers and noises. In addition, the RF algorithm can also evaluate the importance of features, thereby helping to filter the insignificant features to improve simplify the model. An issue regarding RF is that it takes slightly longer to train and can require a lot of memory for very large datasets.

The construction steps of RF model are as follow:

1. Use the Bootstrapping method to randomly extract $n$ samples with replacement from the training set to construct a selected training set, and use random sampling method to select a part of features.

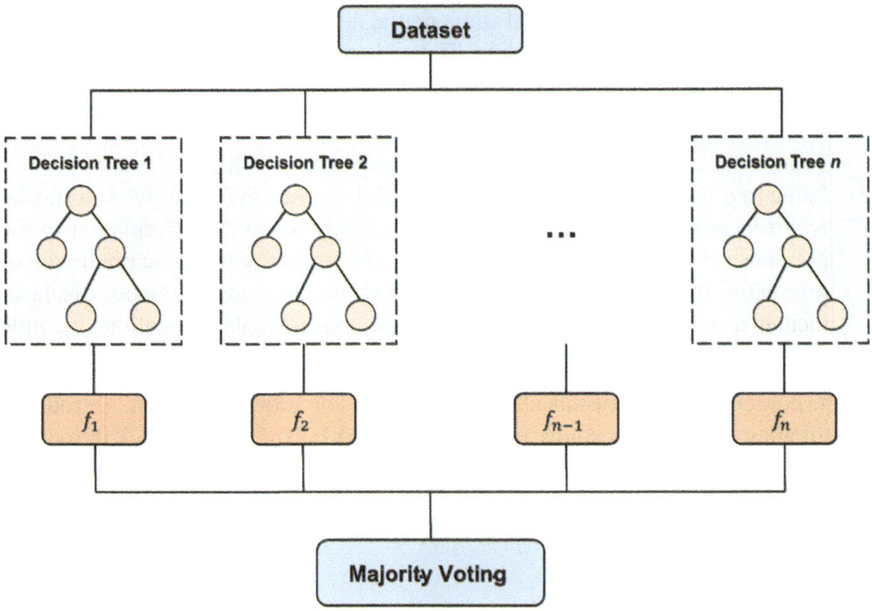


**Fig. 3.1** Principle of random forest algorithm

2. Construct the decision tree based on the selected training set and features.
3. Repeat steps 1 and 2 to generate a series of decision trees.
4. Combine the generated decision trees to form a random forest.
5. Let all decision trees to make predictions for the new input data, and use majority voting to decide the final prediction result.

In general, the advantages of the RF algorithm lie in its high accuracy, fewer parameters, good interpretability, and strong anti-noise ability. These advantages allow RF algorithm to be used in many tasks and scenarios, so the RF model with high generalization ability and classification regression strength is selected as the basic model for the landslide displacement prediction.

### 3.2.2 Long Short-Term Memory Networks

LSTM neural network was firstly proposed by Hochreiter and Schmidhuber (1997). Since then, there have been many studies and articles on the use of LSTM networks for various time series forecasting tasks. It has been proven to be a powerful time series prediction tool. An LSTM network is a type of RNN specifically designed to handle sequential data. RNN is a neural network that can process data with a time dimension, and LSTM network is a variant of RNN that aims to overcome the limitations of traditional RNNs in handling long-term dependencies (Zhang et al., 2021). The structure of LSTM is shown in Fig. 3.2.

LSTM has a few key properties that allow it to be ideal for predicting landslide displacement. Firstly, it works well in processing the data in chronological order. This allows the network to account for the temporal dynamics of landslides, which typically occur over a period of time. In addition, LSTM has a memory mechanism that allows them to maintain a memory of previous inputs. This allows the network to consider the history of displacement measurements and make more accurate predictions. The memory mechanism consists of multiple components, such as forget gate, input gate, and output gate, which work together to selectively store and retrieve information in the network memory. Finally, LSTM is able to handle long-term dependencies, which is a common challenge in time series forecasting problems. Long-term dependencies are data patterns that occur over a long period of time. Traditional RNNs struggle to capture these patterns, but LSTM can achieve this because it maintains the memory of previous inputs. All these properties make LSTM an ideal network for this task.

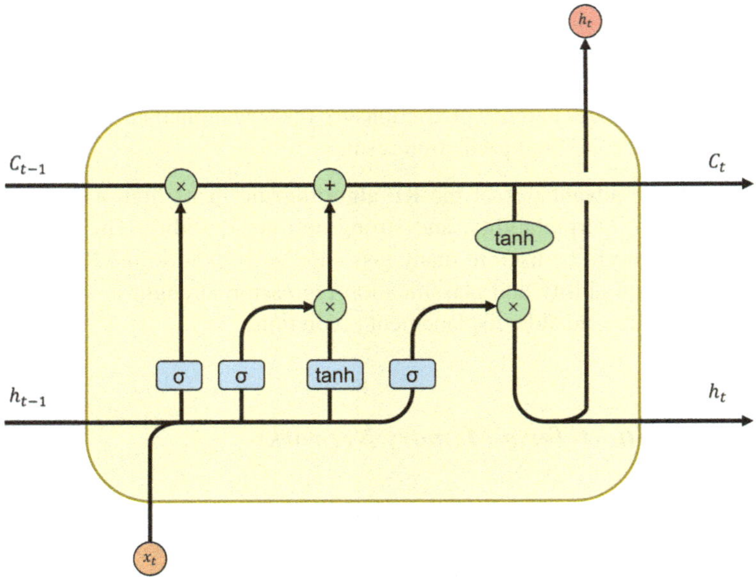


**Fig. 3.2** Principle of long short-term memory neural network

## 3.3 Data Acquisition and Model Construction

### *3.3.1 Data Acquisition*

In this study, the landslide site (31°01′34″ N, 110°32′09″ E, Fig. 3.3) in Baishuihe, Yichang, China, was selected for the landslide displacement study. The Baishuihe landslide is located on the south side of the Yangtze River which is 56 km away from the Three Gorges Dam. It belongs to Baishuihe Village, Shazhenxi Town. The landslide body is located in the wide valley of the Yangtze River. It is a monoclinical bedding slope and high in the south and low in the north, spreading towards the Yangtze River in a ladder shape. The elevation of the rear edge is around 410 m, which is bounded by the rock-soil boundary. The front edge reaches the Yangtze River, and the east and west sides are bounded by bedrock ridges, with an overall slope angle of 30°. Its north–south length is 600 m and east–west width is 700 m. The average thickness of the landslide is about 30 m with the volume of 131,040 $m^3$. It is an accumulative layer landslide with a forward slope (Yi, 2020).

According to the topographical and geological conditions, deformation characteristics, and observation condition of study site, the monitoring contents here was determined to includes surface displacement measurement, borehole survey, and groundwater level monitoring. In the initial layout area of Baishuihe landslide, there were 7 GPS monitoring points, which were distributed in 3 longitudinal sections, of which there are 3 monitoring points (ZG118, ZG119, ZG120) in the middle section,

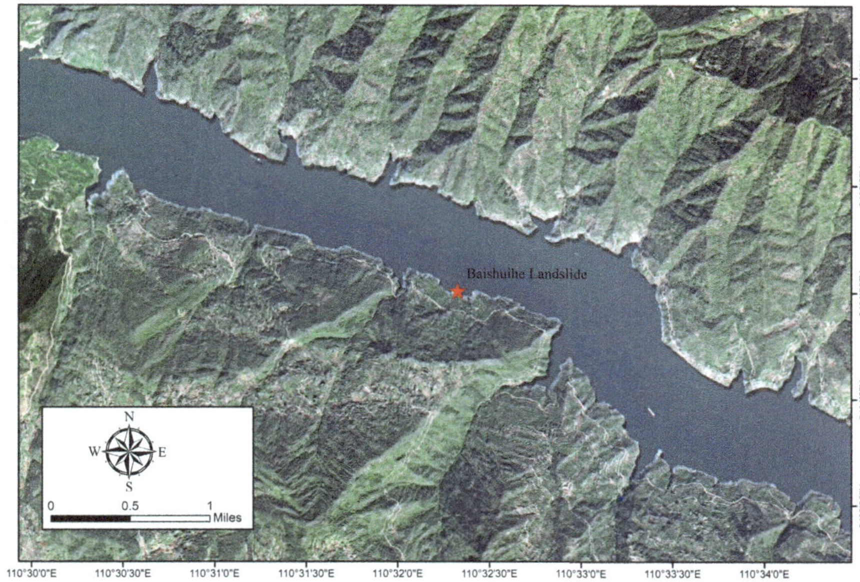


**Fig. 3.3** Study site of Baishuihe landslide

and 2 monitoring points for each side profile (ZG91, ZG94, and ZG92, ZG93). But due to the deformation and cracks, a landslide warning area was then set. Four additional GPS monitoring points (XD-01, XD-02, XD-03, XD-04) were built in the early warning area, and one GPS reference point was set on the bedrock ridges on the east and west sides. So, there are 11 GPS deformation monitoring points in total, and 6 of them are located in the active area of landslide displacement. The U.S. Tembo GPS receiver (plane accuracy 5 + 1 ppm) was used to monitor the surface displacement and deformation of landslide. Detail information regarding the positions of monitoring point can be obtained from the reference (Yi, 2020). Sun et al. (2019) studied the landslide displacement of ZG93 monitoring point before. The ZG118 monitoring point is also located in the landslide early warning area, so the data of ZG118 monitoring point was selected for this analysis. The observation data from 2006 to 2012 were used which was provided by National Cryosphere Desert Data Center, China.

Reservoir water level variation and rainfall are the main external factors leading to landslide displacement. In order to indicate this relationship, we collected data on reservoir water level and rainfall in this area, and the relationship maps are shown in Fig. 3.4. In the summer (rainy) season every year, the landslide deformation increases significantly, and tends to be stable from winter season to spring of the next year, indicating that the reservoir water level fluctuations and rainfall have a significant impact on the stability of the landslide. During the summer time, the rainfall in this area can last for a long time with high intensity, which affects the landslide displacement and make it showing a step change characteristic. Three Gorges Reservoir

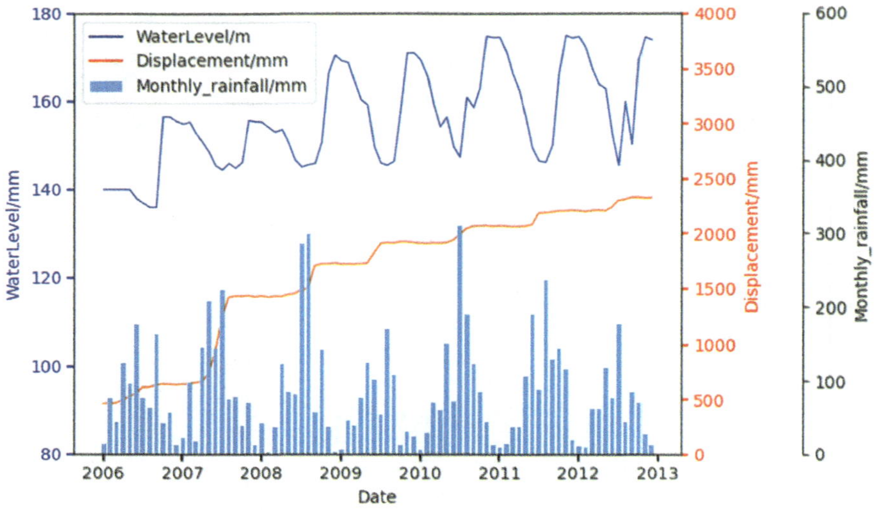


**Fig. 3.4** Relationship of cumulative displacement to reservoir water level and monthly rainfall of ZG118 landslide monitoring point

begins to release water to prevent floods, and the water level of the reservoir drops significantly, indicating negatively relationship with rainfall (Luo et al., 2020). In this case, when the rainfall increases and the water level of the reservoir drops, a sharp change of the landslide displacement can be usually observed.

## *3.3.2 Data Pre-processing*

Data cleaning is particularly important that can help to obvious improve the model performance. Human operation errors, data transmission errors, equipment failures, and ambiguous geological information can affect the original data set. These biased data were removed in advance by using the imputation or elimination of missing values and the identification of outliers on the data.

In addition, different machine learning algorithms have various sensitivity to the dimension difference of input features. It is recommended that before model training, the input features of the training samples should be uniformly normalized or range-scaled to ensure that the sample input feature ranges have no significant difference, otherwise it will directly affect the accuracy of the model.

The change of landslide displacement with time constitutes a displacement time series, and the evolution of landslide displacement is affected by both the characteristics of the landslide and the strength of external factors. The Seasonal and Trend decomposition using Loess (STL) method (Cleveland et al., 1990) was used for the landslide displacement time series decomposition. STL decomposes the time series into three main components: trend component ($T_t$), seasonal component ($S_t$), and residual component ($R_t$):

$$y_t = T_t + S_t + R_t \tag{3.1}$$

Locally estimated scatterplot smoothing (LOESS) is used to extract the three components. In this study, *statsmodels* module of python 3.9 is used to perform the STL.

Exponential smoothing was used to perform the extraction and prediction of trend component of landslide displacement. When doing time series prediction, it is usually believed that the closer the point is to the prediction point, the greater the effect. The weight of the oldest data will be close to 0. Exponential smoothing can be expressed as:

$$S_i = \alpha \sum_{j=0}^{i} (1 - \alpha)^j x_{i-j} \tag{3.2}$$

where $S_i$ is the smoothed value at time step $i$, and $x_i$ is the actual data at this time step. $\alpha$ can be any value between 0 and 1 which controls the balance between old and new information. When it is close to 1, only the current data point is kept. And when it is close to 0, only the previous smooth value is kept. It can be seen that all previous observations contribute to the current smoothed value, but their contribution decreases as the power of $\alpha$ increases. Those relatively earlier observations played a relatively small role. In a way, exponential smoothing is like a moving average with infinite memory and exponentially decreasing weights. The result of an exponential smoothing can be extended to make predictions. The prediction method is:

$$x_{i+h} = S_i \tag{3.3}$$

where $S_i$ is the last calculated value. $h$ equals to 1 represents the following predicted value.

### 3.3.3 Model Construction

Periodic (seasonal) component was divided into train set and test set according to the time. Data from 2006 to 2011 was selected to construct the train set and data after 2011 was used to construct the test set. According to the analysis of Fig. 3.4, when the rainfall increases and the water level of the reservoir drops during the summer time, change of the landslide displacement is more dramatic than other seasons. These two external factors were included to study the landslide displacement. Considering the hysteresis of the two factors on landslides, the two factors were processed to the forms of monthly mean water level of the reservoir, monthly reservoir water level change amplitude, monthly rainfall and cumulative rainfall in the previous 3 months.

To study the displacement prediction of Baishuihe landslide, the *RandomForestRegressor* class of the *scikit-learn* library in Python 3.9 is used to construct the RF model for landslide displacement prediction in this study. The Bayesian optimization algorithm was used to search for the optimal hyperparameter of the RF model. The LSTM model was constructed by using the *Keras* module in Python 3.9. The model consists of input layer, LSTM hidden layer and output layer. The number of neurons in the input layer is determined to be 4 according to the number of environmental factors abovementioned. The hidden layer used a grid search method to find optimal hyperparameters. Since only the displacement needed to be predicted, the number of neurons in the output layer was set to be one.

We evaluated the performance of RF and LSTM network by using several evaluation metrics to measure their predictive accuracy. The coefficient of determination ($R^2$) is used to indicate how well the machine learning model make prediction. The mean absolute error (MAE) is a measure of the difference between predicted and actual values, with lower values indicating a better fit. Root mean square error (RMSE) is the average of forecast errors. The calculation formulas of $R^2$, MAE and RMAE can be expressed as:

$$R^2 = \left( \frac{\sum_{i=1}^{n} (X_i - \bar{X})(Y_i - \bar{Y})}{\sqrt{\sum_{i=1}^{n} (X_i - \bar{X})^2} \sqrt{\sum_{i=1}^{n} (Y_i - \bar{Y})^2}} \right)^2 \tag{3.4}$$

$$MAE = \frac{1}{n} \sum_{i=1}^{n} (X - Y) \tag{3.5}$$

$$RMSE = \sqrt{\frac{1}{n} \sum_{i=1}^{n} (X - Y)^2} \tag{3.6}$$

where $n$ is the number of pairs of predicted and measured landslide displacement, $X$ and $Y$ are the predicted and measured landslide displacement.

## 3.4 Results and Analysis

### 3.4.1 Prediction of Trend Landslide Displacements

Comparison of prediction and measurement of trend landslide displacement in ZG118 landslide monitoring point from 2006 to 2012 is indicated in Fig. 3.5. The performance of one-dimensional exponential smoothing is of high quality considering the variations of trend displacement are almost same for the predicted and measured values.

### 3.4.2 Prediction of Periodic Landslide Displacements

In the optimization results of RF model, the number of constructed decision trees (*n_estimators*) was 167, the maximum depth of decision trees (*max_depths*) was 10, and the minimum number of samples of decision tree nodes (*min_samples_split*) was 4. If the number of samples of a node is < 4, it will not be split any more. Other parameters adopted the default values of the RF module. Regarding the LSTM model, the finally determined number of hidden layers is one and the number of neurons is 16.

Prediction by RF and LSTM and measurement of periodic displacements at ZG118 landslide monitoring point from 2006 to 2013 are indicated in Figs. 3.6 and 3.7, and the results of performance metrics are shown in Table 3.1. Overall, both models successfully predicted the periodic displacements. The predicted variation

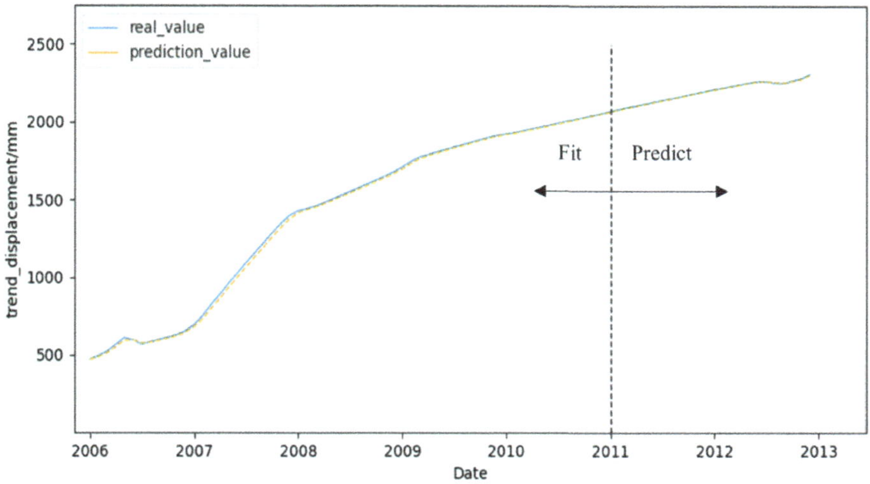


**Fig. 3.5** Prediction and measurement of trend displacements at ZG118 landslide monitoring point from 2006 to 2013

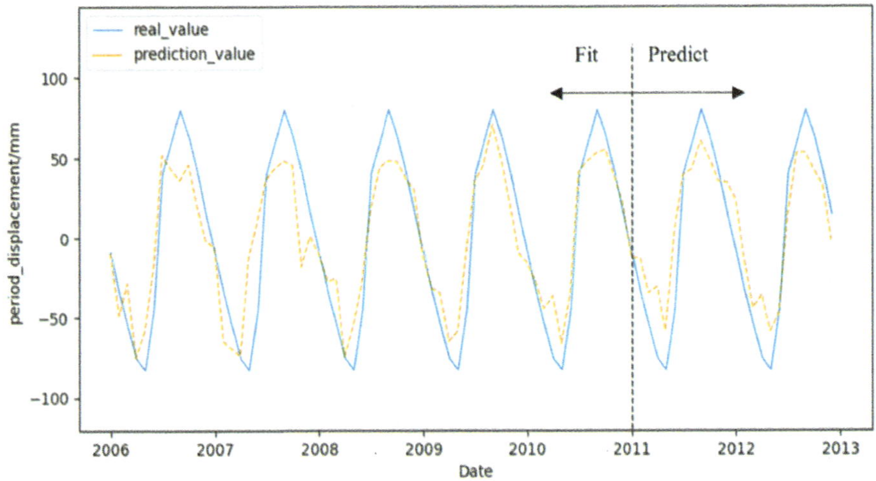


**Fig. 3.6** RF Prediction and measurement of periodic displacements at ZG118 landslide monitoring point from 2006 to 2013

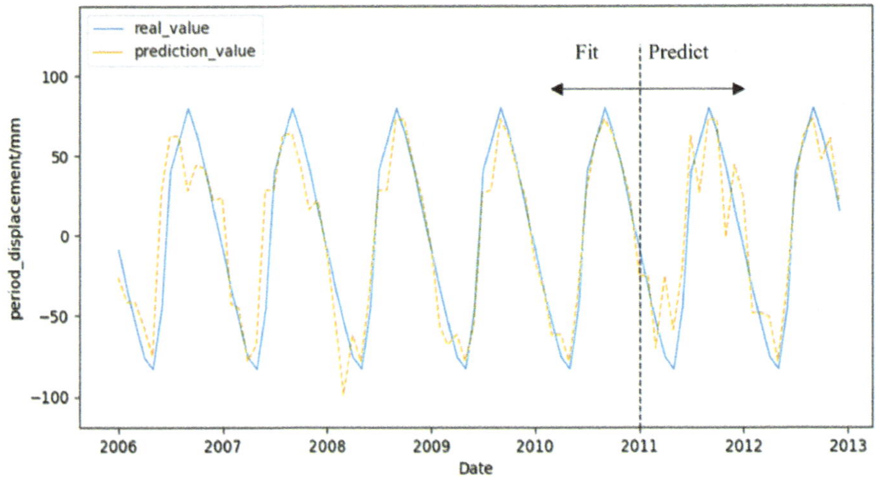


**Fig. 3.7** LSTM Prediction and measurement of periodic displacements at ZG118 landslide monitoring point from 2006 to 2013

trend of periodic displacements is close to that of the measured one. But for the peak value and bottom value, the RF cannot work well. In contrast, the LSTM model to some extent solve the problem with a relative better performance for the edge values.

From the evaluation metrics in Table 3.1, LSTM model indicate better performance with $R^2$ of 0.84, MAE of 18.38 mm and RMSE of 22.01 mm, so the prediction

**Table 3.1** Performance of RF and LSTM in predicting the periodic displacement at ZG118 landslide monitoring point

| | RF | LSTM |
|---|---|---|
| | 23.67 mm | 22.01 mm |
| MAE | 19.99 mm | 18.38 mm |
| $R^2$ | 0.82 | 0.84 |

by LSTM is relatively closer to most of the measured values, and can better indicate the variation of the periodic landslide displacement.

### 3.4.3 *Prediction of Cumulated Landslide Displacements*

The predicted trend displacement and periodic displacement were finally combined to obtain the predicted value of cumulative landslide displacement. Prediction by RF and LSTM and measurement of cumulated displacements at ZG118 landslide monitoring point from 2006 to 2013 are shown in Figs. 3.8 and 3.9, and the values of performance metrics are shown in Table 3.2. It is astonishing that RF model indicate a little bit better performance in cumulated displacement prediction, with the $R^2$ of 0.95, MAE of 16.56 mm and RMSE of 19.85 mm. It can be seen that the predicted value of the RF model is closer to the measured value during the stable months with minor displacement variations. As the number of months with minor displacement variation is more than the months with significant displacement variation, RF model indicate the better performance. Overall, both models can well reveal the characteristics of the Baishuihe landslide's intermittent activity.

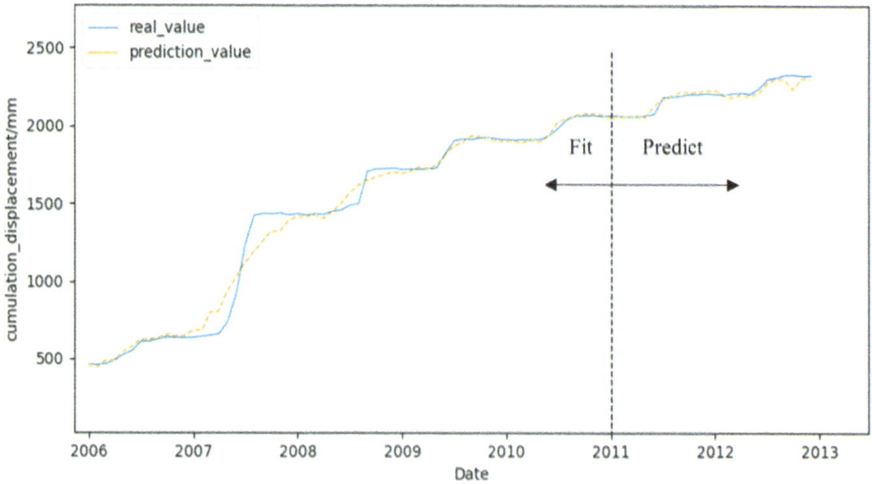


**Fig. 3.8** RF Prediction and measurement of cumulated displacements at ZG118 landslide monitoring point from 2006 to 2013

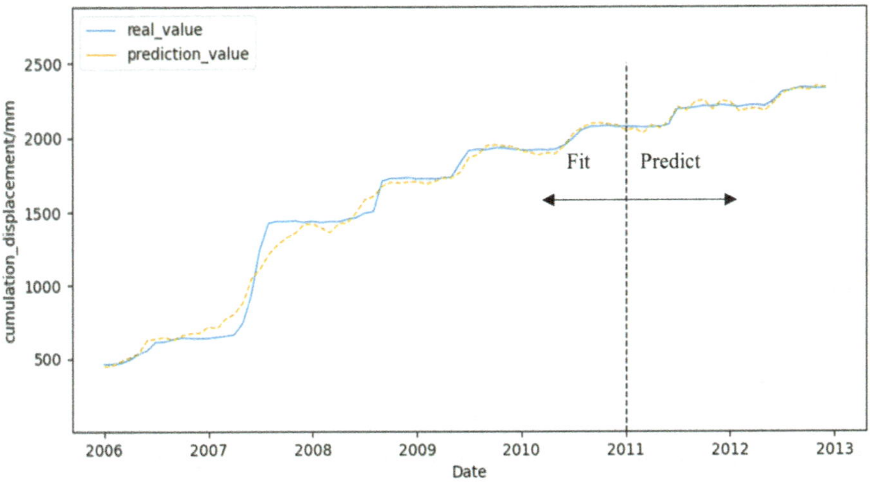


**Fig. 3.9** LSTM Prediction and measurement of cumulated displacements at ZG118 landslide monitoring point from 2006 to 2013

**Table 3.2** Performance of RF and LSTM in predicting the cumulated displacement at ZG118 landslide monitoring point

| | RF | LSTM |
|---|---|---|
| RMSE | 19.85 mm | 23.33 mm |
| MAE | 16.56 mm | 20.77 mm |
| $R^2$ | 0.95 | 0.94 |

## 3.5 Summary

This chapter proposed a RF and LSTM model to study the long-term landslide displacement based on multiple environmental factors. Ground data of Baishuihe landslide including measured displacement, reservoir water level and rainfall record were considered in this study to construct the landslide displacement dataset with multi-source long-term monitoring data. Using STL can accurately separate the time series landslide displacement to trend displacement, periodic displacement and residual displacement. Exponential smoothing successfully predicted the trend displacement. LSTM model performed better in the prediction of periodic displacement, but did worse in the prediction of cumulated displacement, while the RF model is opposite. Both models can be used in this task. These results demonstrate the effectiveness of using machine learning methods for landslide displacement prediction and its potential to improve early warning and risk reduction in landslide vulnerable regions.

## Codes

## [Python3.9] Random Forest Regression

```
import sys
import pandas as pd
import numpy as np
from sklearn.ensemble import RandomForestRegressor
from sklearn.metrics import confusion_matrix
import matplotlib.pyplot as plt
from sklearn.model_selection import train_test_split
from sklearn.model_selection import RandomizedSearchCV
from sklearn.metrics import mean_squared_error, explained_
variance_score, mean_absolute_error, r2_score
data = pd.read_csv('./air_train&test.csv',index_col = 0,encoding
= 'gb2312')
print (data.head())
print (data.shape)
index = data.index
col = data.columns
class_names = np.unique(data.iloc[:,-1])
#print (type(data))
print (class_names)
#print (data.describe())
#Split train set and test set
data_train, data_test = train_test_split(data,test_size = 0.1,
random_state = 0)
X_train = data_train.iloc[:,0:-2]
X_test = data_test.iloc[:,0:-2]
feature = data_train.iloc[:,0:-2].columns
print (feature)
y_train = data_train.iloc[:,-2]
y_test = data_test.iloc[:,-2]
data.drop([u'Landslide Level'],axis = 1).corr()
import seaborn as sns
sns.set(style = "ticks", color_codes = True);
palette = sns.xkcd_palette(['dark blue', 'dark green', 'gold',
'orange'])
sns.pairplot(data.drop([u' Landslide Level '],axis = 1), diag_kind
= 'kde', plot_kws = dict(alpha = 0.7))
plt.show()
#parameter tunning
criterion = ['mse','mae']
n_estimators = [int(x) for x in np.linspace(start = 200, stop =
2000, num = 10)]
max_features = ['auto', 'sqrt']
max_depth = [int(x) for x in np.linspace(10, 100, num = 10)]
max_depth.append(None)
min_samples_split = [2, 5, 10]
min_samples_leaf = [1, 2, 4]
bootstrap = [True, False]
random_grid = {'criterion':criterion,
         'n_estimators': n_estimators,
```

```
        'max_features': max_features,
        'max_depth': max_depth,
        'min_samples_split': min_samples_split,
        'min_samples_leaf': min_samples_leaf,
        'bootstrap': bootstrap}
clf = RandomForestRegressor()
clf_random = RandomizedSearchCV(estimator = clf, param_
distributions = random_grid,
                 n_iter = 10, cv = 3, verbose = 2, random_state =
42, n_jobs = 1)
clf_random.fit(X_train, y_train)
print (clf_random.best_params_)
print ('r2:', rf.score(X_test,y_test))
print (' RMSE ', math.sqrt(mean_squared_error(y_test,y_test_
pred)))
print ('MAE', mean_absolute_error(y_test,y_test_pred))
rf.feature_importances_#[rf.feature_importances_ > 0.01]
#prediction
data_pred = pd.read_csv('./air.csv',index_col = 0,encoding =
'gb2312')
index = data_pred.index
y_pred = rf.predict(data_pred.values)
```

**[Python3.9] LSTM Regression**

```
import numpy as np
import pandas as pd
from keras.callbacks import ModelCheckpoint
from sklearn.preprocessing import MinMaxScaler
from sklearn.preprocessing import LabelEncoder
from keras.models import Sequential
from keras.layers import Dense, LSTM, GRU, Dropout
from numpy import concatenate
from sklearn.metrics import mean_squared_error, r2_score, mean_
absolute_error
from scipy import interpolate
import math
import matplotlib.pyplot as plt
#lookback_window
def data_split(data, train_len, lookback_window):
  train = data[:train_len]
  test = data[train_len:]
  # print(train.shape)
  X1, Y1 = [], []
  for i in range(lookback_window, len(train)):
    X1.append(train[i - lookback_window:i])
    Y1.append(train[i])
  Y_train = np.array(Y1)
  X_train = np.array(X1)
  X2, Y2 = [], []
  for i in range(lookback_window, len(test)):
    X2.append(test[i - lookback_window:i])
    Y2.append(test[i])
  Y_test = np.array(Y2)
  X_test = np.array(X2)
  print(X_train.shape)
```

```
    print(Y_train.shape)
    return (X_train, Y_train, X_test, Y_test)
def data_split_LSTM(X_train,Y_train, X_test, Y_test):
    X_train = X_train.reshape(X_train.shape[0], X_train.shape[1],
1)
    X_test = X_test.reshape(X_test.shape[0], X_test.shape[1], 1)
    Y_train = Y_train.reshape(Y_train.shape[0], 1)
    Y_test = Y_test.reshape(Y_test.shape[0], 1)
    return (X_train, Y_train, X_test, Y_test)
def imf_data(data, lookback_window):
    X1 = []
    for i in range(lookback_window, len(data)):
        X1.append(data[i - lookback_window:i])
    X1.append(data[len(data)-1:len(data)])
    X_train = np.array(X1)
    return (X_train)
def visualize(history):
    plt.rcParams['figure.figsize'] = (10.0, 6.0)
    # Plot training & validation loss values
    plt.plot(history.history['loss'])
    plt.plot(history.history['val_loss'])
    plt.title('Model loss')
    plt.ylabel('Loss')
    plt.xlabel('Epoch')
    plt.legend(['Train', 'Test'], loc = 'upper left')
    plt.show()
def LSTM_Model(X_train, Y_train):
    filepath = '../LSTM/LSTM-{epoch:02d}.h5'
    checkpoint = ModelCheckpoint(filepath,
                        monitor = 'loss',
                        verbose = 1,
                        save_best_only = False,
                        mode = 'auto',
                            period = 10)
    callbacks_list = [checkpoint]
    model = Sequential()
        model.add(LSTM(50,  input_shape  =  (X_train.shape[1],  X_
train.shape[2])))
    model.add(Dense(1))
    model.compile(loss = 'mse', optimizer = 'adam')
    his = model.fit(X_train, Y_train, epochs = 100, batch_size = 16,
validation_split = 0.1, verbose = 2, shuffle = True)
    return (model,his)
# Plot variation curve
def plot_curve(true_data, predicted):
    plt.plot(true_data, label = 'True data')
    plt.plot(predicted, label = 'Predicted data')
    print(true_data)
    print(predicted)
    rmse = format(RMSE(true_data, predicted), '0.4f')
    mape = format(MAPE(true_data, predicted), '0.4f')
    r2 = format(r2_score(true_data, predicted), '0.4f')
    mae = format(mean_absolute_error(true_data, predicted), '0.4f')
```

```
    print('RMSE:' + str(rmse) + '\n' + 'MAE:' + str(mae) + '\n' +
'MAPE:' + str(mape) + '\n' + 'R2:' + str(r2))
    plt.legend()
    # plt.plot(predicted_LSTM, label = 'Predicted data by LSTM')
plt.legend()
    plt.savefig('result_final.png')
    plt.show()
def RMSE(test, predicted):
    rmse = math.sqrt(mean_squared_error(test, predicted))
    return rmse
if __name__ = = '__main__':
    plt.rcParams['figure.figsize'] = (10.0, 5.0) # set default size
of plots
    plt.rcParams['image.interpolation'] = 'nearest'
    plt.rcParams['image.cmap'] = 'gray'
    dataset = pd.read_csv('../csv/air_train&test1.csv', header = 0,
index_col = 0, parse_dates = True)
    data = dataset.values.reshape(-1)
    values = dataset.values
    # groups = [0, 1, 2, 3]
    # fig, axs = plt.subplots(1)
    df = pd.DataFrame(dataset)
    do = df['Dissolved Oxygen']
    DO = []
    for i in range(0, len(do)):
        DO.append([do[i]])
    scaler_DO = MinMaxScaler(feature_range = (0, 1))
    DO = scaler_DO.fit_transform(DO)
    # plt.plot(DO)
    c = int(len(df) * 0.8)
    lookback_window = 10
    X1_train, Y1_train, X1_test, Y1_test = data_split(DO, c, look-
back_window) # TCN
    X2_train, Y2_train, X2_test, Y2_test = data_split_LSTM(X1_train,
Y1_train, X1_test, Y1_test)
    # Model training
    model_DO_LSTM,his = LSTM_Model(X2_train, Y2_train)
    visualize(history)
    #Testing
    Y2_train_hat = model_DO_LSTM.predict(X2_train)
    Y2_train_hat = scaler_DO.inverse_transform(Y2_train_hat)
    Y2_train = scaler_DO.inverse_transform(Y2_train)
    print(Y2_train.ndim)
    print(Y2_train_hat.ndim)
    Y2_test_hat = model_DO_LSTM.predict(X2_test)
    test = Y2_test
    prediction = Y2_test_hat
    Y2_test = scaler_DO.inverse_transform(Y2_test)
    Y2_test_hat = scaler_DO.inverse_transform(Y2_test_hat)
    plot_curve(Y2_train, Y2_train_hat)
    plot_curve(Y2_test, Y2_test_hat)
    rmse = format(RMSE(test, prediction), '0.4f')
    r2 = format(r2_score(test, prediction), '0.4f')
    mae = format(mean_absolute_error(test, prediction), '0.4f')
```

## References

Breiman, L. (2001). Random forests. *Machine Learning, 45*, 5–32.

Cleveland, R. B., Cleveland, W. S., McRae, J. E., & Terpenning, I. (1990). STL: A seasonal-trend decomposition. *Journal of Official Statistics, 6*, 3–73.

Gers, F. A., Schraudolph, N. N., & Schmidhuber, J. (2002). Learning precise timing with LSTM recurrent networks. *Journal of Machine Learning Research, 3*, 115–143.

Hochreiter, S., & Schmidhuber, J. (1997). Long short-term memory. *Neural Computation, 9*, 1735–1780.

Liang, Y., Xiao, T., Hu, C., Ren, S., & Zeng, L. (2022). Prediction of landslide displacement based on long term monitoring data and LSTM network. *Journal of Signal Processing, 38*, 19–27.

Luo, H., Jiang, Y., Xu, Q., & Tang, B. (2020). Time series analysis and combined modeling prediction of GPS landslide displacement monitoring. *Journal of Institute of Disaster Prevention, 22*, 20–28.

Sun, Y., Li, D., Yin, K., Chen, L., & Wang, Y. (2019). Intermittent movement prediction of colluvial landslide in the three gorges reservoir: A case study of Baishuihe landslide. *Geological Science and Technology Information, 38*(5), 195–203.

Wang, C., Li, L., Wen, Z., Zhang, M., & Wei, X. (2022). Dynamic prediction of landslide displacement based on time series and CNN-LSTM. *Foreign Electronic Measurement Technology, 41*, 1–8.

Xie, M., Ju, N., Zhao, J., Fan, Q., & He, C. (2021). Comparative analysis on classification methods of geological disaster susceptibility assessment. *Geomatics and Information Science of Wuhan University, 46*(7), 1003–1014.

Yan, H., Chen, J., Li, S., & Wu, L. (2021). Predicting of landslide displacement based on time series and Gated Recurrent Unit. *Yangtze River, 158*, 102–107.

Yi, W. (2020). *Basic characteristics and monitoring data of Baishuihe landslide in Zigui County, Three Gorges Reservoir area, 2007–2012*. National Cryosphere Desert Data Center.

Zhang, M., Li, L., & Wen, Z. (2021). Research on RNN and LSTM method for dynamic prediction of landslide displacement. *Pearl River, 42*, 6–13.

Zhou, C., Yin, K., & Huang, F. (2015). Application of the chaotic sequence WA-ELM coupling model in landslide displacement prediction. *Rock and Soil Mechanics, 36*, 2674–2680.

Zhou, P., Deng, H., Zhang, W., Xue, D., Wu, X., & Zhuo, W. (2022). Landslide susceptibility evaluation based on information value model and machine learning method: A case study of Lixian County, Sichuan Province. *Scientia Geographica Sinica, 42*, 1665–1675.

# Chapter 4
# Deep Learning for Long-Term Landslide Change Detection from Optical Remote Sensing Data

**Abstract** Landslides are widespread in China, but are only mapped in detail in limited scales and within a short period of time. In this study, long-term Landsat images and DEM were integrated into machine learning models to conduct long-term dynamic monitoring of landslides in the severely affected area of the Wenchuan Earthquake. The study found that the U-Net model can achieve high detection accuracy and the integration multi-source data helps the model to well understand the landslides and significantly improves the model performance. The model also shows a fairly high accuracy in non-training years, indicating the good time invariance in landslide identification. This study contributes to the development of remote sensing time series analysis technology and landslide remote sensing identification method, and also help to promote the extensive development of historical landslide cataloging and long-term landslide activity analysis, which will better support the major engineering construction and mountainous land development and management.

## 4.1 Introduction

### *4.1.1 Background and Significance*

China is a country susceptible to landslides and other geological disasters due to active tectonics, complex geological environment, diverse climate types, and frequent human engineering activities (Liu & Chen, 2020; Zhang et al., 2013). The geological disaster notification data indicated that a total of 409,000 geological disasters occurred across the country from 2000 to 2020, of which 293,000 were landslides, accounting for 71.6%. More than 12,000 people died or disappeared due to geological disasters, and the direct economic loss exceeded 80 billion RMB. In order to further explore the development and evolution mechanism of landslides for the sensitivity, hazard and risk analysis and early warning, it is firstly necessary to perform landslide identification and cataloging (Assilzadeh et al., 2010; Brardinoni et al., 2003; Guzzetti et al., 2012).

The history of landslides records the occurrence and development of landslides over a long period of time, and is an important basis for analyzing the spatial

 
W. Chen et al., *Intelligent Interpretation for Geological Disasters*,
https://doi.org/10.1007/978-981-99-5822-1_4

and temporal patterns of regional landslide activity and assessing regional stability (Althuwaynee et al., 2012). The obtained regional cumulative incidence of landslides has been widely adopted as an important indicator for landslide prediction (Chuang & Shiu, 2018). Other time-dimensional indicators such as landslide duration and recurrence rate can also be used to study landslide development conditions such as slip surface depth, rock and soil properties, and hydrogeological characteristics (Behling et al., 2014). Historical catalog data can be used to analyze and predict long-term landslide activity, which is of great importance for guiding major projects and urban construction planning in mountainous areas (Chen et al., 2019).

Studies have shown that a large number of loose accumulations formed by coseismic geological disasters induced by large earthquakes will continue to affect the evolution of watersheds and regional stability (Huang & Li, 2014; Wang et al., 2017). Earthquake-affected areas usually have to face the threat of landslides and debris flows for decades (Fan et al., 2019; Li et al., 2018). In light of this, in major project (such as the Three Gorges Project, Sichuan-Tibet Railway) construction sites and extreme natural event (such as earthquakes, extreme precipitation) susceptible areas, long-term landslide monitoring and construction of fine long-term landslide historical dynamic library will help prevent major regional disasters, protect the safety of people's lives and property and the ecological environment, and maintain national security and social stability.

### 4.1.2 Research Overview

At present, the historical dynamics of landslides are very lacking in China and even in most regions of the world (Chen et al., 2019). Due to the high cost of high-resolution remote sensing data, and the workload and difficulty of landslide interpretation are huge, and the continuous landslide cataloging project is only carried out in limited regions. In most regions there were only landslide mapping for work or research needs, and it is difficult to ensure that the landslide catalog can be updated annually (Xu et al., 2014). If the update interval of historical landslide mapping is too long, it would be hard to completely record the development and evolution process of landslides, and the landslides that occurred during this interval will be overlooked. However, for regions lacking the historical dynamics of landslides, the early landslides and their entire evolution process cannot be well revealed. These problems have restricted the researches on early regional landslide activity, development conditions and evolution mechanism, and brought greater uncertainty to regional stability assessment (Du et al., 2020; Wang et al., 2020).

In recent years, with the development of satellite remote sensing and the accumulation of historical data, a large amount of time series remote sensing data has been formed (Du et al., 2019; Zhao et al., 2020). The Landsat series satellites have accumulated nearly 50 years of time-series images so far, objectively recording the dynamic changes of the earth's surface in a long-term range, which can be used for the study of the spatio-temporal dynamic variations of ground objects, especially the evolution process of landslides (Chen et al., 2019; Mwaniki et al., 2017). Many studies have shown that Landsat series and Sentinel-2 images are sufficient for most regional landslide mapping tasks (Mwaniki et al., 2015; Zhong et al., 2021).

In terms of specific algorithms, a variety of knowledge-driven and data-driven methods are used for classification and recognition, such as analytic hierarchy process, decision tree (DT), artificial neural network (ANN), support vector machine (SVM) and random forest (RF), etc. (Hölbling et al., 2012; Mezaal et al., 2018). In recent years, deep learning methods, which have achieved great success in the fields of computer vision and image processing, have also been used in landslide recognition research (Cheng et al., 2021; Fang et al., 2020). Sameen and Pradhan (2019) fused imagery and terrain data to train residual networks (ResNets) and the obtained F1-scores and mIoUs that were 13% and 12.96% higher than other models, respectively. Yi and Zhang (2020) proposed a cascaded end-to-end deep learning network (LandsNet), which learns various characteristics of landslides by using automatically generated training samples, and the best F1-Score obtained is about 0.86. Ghorbanzadeh et al. (2019) compared and evaluated the performance of ANN, SVM, RF and convolution neural network (CNN). The CNN with small window size achieved the best result of 78.26% mIOU. Li et al. (2021) used large-scale landslide samples to compare the performance of several typical deep learning models (VGG, ResNet, DenseNet, and U-Net) in identifying landslides in the Wenchuan earthquake-stricken area. Ju et al. (2020) collected 2498 samples of loess landslides using Google images, and then used Mask Region Convolutional Neural Network (Mask R-CNN) for automatic identification of loess landslides with a recall of 0.72 and an F1-score of 0.63 finally obtained. Jiang et al. (2021) proposed to use landslide shape, color, texture and other features to simulate a more complex landslide background to enhance the difficult samples and integrate them with Mask R-CNN network for fine landslide detection, achieving a detection accuracy of 94.0%.

In general, the landslide identification method based on single-temporal images is difficult to distinguish landslides from bare soil, rocks, roads that have similar spectral features, which leads to the high false alarm rate of landslides. And the change detection method based on a small number of time series data may wrongly regards the cultivated land harvesting, engineering construction and seasonal changes in vegetation as landslides (Chen et al., 2018; Lv et al., 2018). To overcome these problems, some studies considered the time series characteristics of NDVI variation in landslide areas to set NDVI thresholds at certain time nodes to identify landslides (Hu et al., 2018). The thresholds are mainly determined based on local conditions and expert experience (Decuyper et al., 2022), which has certain subjectivity and limitations. When the data, research area or time changes, it often needs to be re-tested and adjusted for generalization. But actually, due to the influence of various factors, there may be many cases that do not meet the standard curve and threshold requirements (Yang et al., 2018).

In addition, most of the current studies are carried out on landslide mapping in a single phase. The abovementioned algorithms may achieve better performance in a single phase, but when long-term landslide mapping is required, which method has better temporal coherence and can better overcome the issue of error accumulation is still lacking in our understanding. Chen et al. (2019) used the RF algorithm to carry out the historical dynamics of landslides in Taiwan, but due to the lack of comparison, it is difficult to evaluate the performance of the algorithm horizontally. Also, these methods need to carry out sample collection, model training and classification annually, leading to a heavy workload and low efficiency.

### *4.1.3 Research Object and Contents*

This chapter is to find out a classification algorithm that is not strongly dependent on the samples, thereby reducing the workload of annual sampling. To achieve this, this chapter used nearly 10 years landslide history data in the core disaster area of Wenchuan earthquake to carry out experiments on typical deep and shallow machine learning landslide identification methods, such as RF and U-Net, to figure out which algorithm can have better spatio-temporal coherence to avoid the accumulation of errors, and if it is possible to carry out long-term landslide cataloging using only a few annual samples.

The arrangement of each section is: In Sect. 4.2, the condition of study site and data used in this study are introduced. Section 4.3 describes the principle and construction process of the landslide recognizing models based on RF and U-Net, data sampling procedure, and model evaluation methods. Section 4.4 describes the results of landslide recognizing models and makes comparison on their performance. Sections 4.5 and 4.6 discusses and summarizes the research and illustrates the perspectives in future researches.

## 4.2 Study Area and Dataset

### *4.2.1 Study Area*

The study area (Fig. 4.1) is located in the Longmenshan fault zone in the Wenchuan earthquake-stricken area, which is one of the most seriously affected areas by the Wenchuan earthquake. The Wenchuan earthquake triggered nearly 200,000 coseismic landslides (Xu et al., 2014), distributed over an area of 110,000 $km^2$. After earthquakes, rainfall-induced coseismic debris accumulation failures occur frequently, often turning into catastrophic debris flows and floods (Fan et al., 2019; Huang & Li, 2014; Li et al., 2018; Wang et al., 2017). The study area covers 42 catchment areas from Yingxiu Town (the epicenter) to Wenchuan Town, with an area of more than 471 $km^2$. Coseismic landslides, post-seismic reactivation, and new faults have resulted in multiple periods of landslides in this area, with a total volume between 0.8 and 1.5 billion $m^3$ (Parker et al., 2011; Xu et al., 2016).

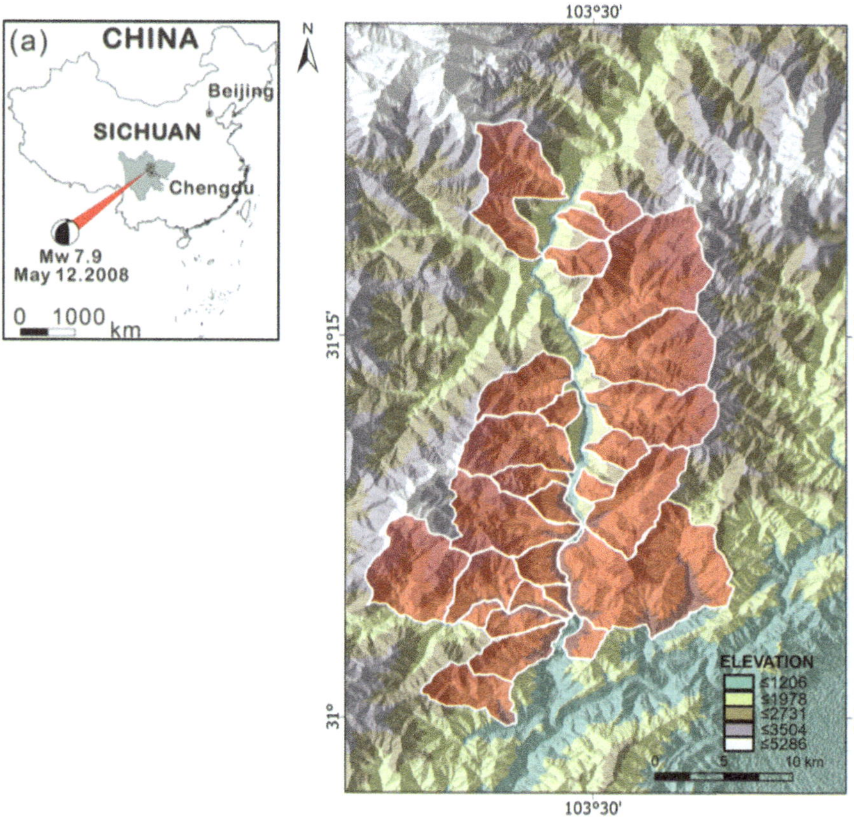


**Fig. 4.1** Study site in Wenchuan, China, modified from Fan et al. (2019)

The terrain is rugged and steep, with elevations ranging from 420 m in the river valley to 6100 m in the Hengduan Mountains. Geological structure and stratigraphic strikes show a general northeast to southwest orientation, with bedrock exposures being highly fractured and weathered. Most low-order channels are deeply cut and steeply sloped, and their morphology is strongly controlled by high tectonic activities. More than half of the slope is steeper than 36°. The study area is surrounded by two major faults in the Longmen Mountains (Wenchuan-Maowen and Yingxiu-Beichuan faults). Weathered and highly fractured rocks are covered by the dense vegetation and soil layers with varying thickness. These rocks are mainly igneous rocks (granite, diorite), a small part of metamorphic and sedimentary rocks (schist, shale, sandstone, limestone). The climate in the study area belongs to the subtropical climate, affected by the monsoon circulation. The annual average temperature is 13 °C, and the annual rainfall is greater than 1250 mm, mainly in summer (Guo et al., 2016).

Complex geology and precipitation patterns (frequent but localized torrential rains that can bring hundreds of millimeters of rain each time) make the region highly prone to debris flows. About 250 debris flows were recorded in the decades before the

Wenchuan earthquake (Cui et al., 2008), and hundreds of debris flows were triggered after the earthquake, affecting more than 800 rivers in the first two years (Cui et al., 2011). The characteristics of rainfall events that trigger debris flows change abruptly with earthquakes, with a marked decrease in the intensity and duration, followed by a gradual recovery pattern over the following decade. Since 2008, researchers have carried out time-series landslide interpretation work in this area for more than 10 years, with comprehensive coverage and high precision, providing a high-quality ground reference for this study.

## *4.2.2 Available Data*

### 4.2.2.1 Time Series Landsat Images

This study collects all available level 2 Landsat-5 and Landsat-8 images covering the study area between 2008 and 2018 from Google Earth Engine (GEE). Level 2 products are surface reflectance data with a spatial resolution of 30 m, systematically processed with terrain and atmospheric corrections, and provided with per-pixel quality information, including cloud occlusion information generated with CFMask (Foga et al., 2017). Due to the failure of the scan line corrector of the Landsat-7 ETM+ satellite (called SLC shutdown), missing bands appeared in the image data, which made data analysis more difficult (Markham et al., 2004). Therefore, ETM+ images were excluded in this experiment. According to the time coverage of Landsat-5 and Landsat-8, there is no Landsat image available in 2012 in this study.

In terms of bands, blue, green, red, near-infrared, first and second short-wave infrared bands of Landsat images were included in this study. Landsat mass bands were used to detect and remove cloud pixels with medium to high confidence. Specifically, the CFMask cloud occlusion algorithm was used to detect and remove cloud pixels on all Landsat images. Then, median synthesis was performed on the removed pixels to form a cloud-free Landsat annual composite image.

### 4.2.2.2 Time Series Landslide Cataloging

The time series landslide catalog is mainly used as ground reference and training samples for the detection of long time series landslide changes. In this study, the time series landslide catalog is derived from the long-term landslide catalog released by Fan et al. (2019). Data was obtained by visual interpretation of high-resolution aerial and satellite imagery (Spot 5, Spot 6, Worldview 2, Pleiades). Rigorous consideration was given to image availability, date of acquisition, coverage, cloud cover, and resolution during interpretation. The coverage of remote sensing images in the study area was close to 99% in 2007, 2011 and 2015, 97% in 2008, and 95% in 2013.

The dataset contains annual landslide catalogs for 2005, 2007, 2008, 2011, 2013, 2015, 2017, and 2018 in the study area. On this basis, the visual interpretation was performed in this study of Google images to carry out supplementary work on the data of other years, and completed the annual landslide catalogs in 2009, 2010, 2014, and 2016, forming a time series landslide catalog of more than 10 years. It provides a complete ground reference for long-term landslide dynamic change detection.

#### 4.2.2.3 Digital Elevation Model

The 30 m DEM Shuttle Radar Topography Mission (SRTM) (ftp://e0mss21u.ecs.nasa.gov/srtm/) generated in the collaboration between NASA and the National Geospatial-Intelligence Agency was also used. In this study, the data was used to obtain information including the elevation, slope, and aspect of the landslide for landslide identification. This study attempted to improve the quality of landslide identification by fusing imagery and terrain information.

The SRTM data are located between 60° north latitude and 56° south latitude, covering an area of more than 119 million km$^2$, which is more than 80% of the global land surface. The raw SRTM elevation data was processed at 1 arcsecond intervals from c-band radar signals at NASA's Jet Propulsion Laboratory (JPL). Raw data for regions outside the United States are publicly released at 3 arcseconds and are known as SRTM3 data. SRTM1 (1 arcsecond) contains height data of 3601 $\times$ 3601 sampling points, with a resolution of about 30 m. SRTM3 (3 arcseconds) contains elevation data of 1201 $\times$ 1201 sampling points, with a resolution of about 90 m. SRTM plane accuracy is $\pm$ 20 m, and elevation accuracy is $\pm$ 16 m. In addition, using the SRTM radar data of repeated orbits can achieve the differential interferometry, which can be used to monitor crustal deformation and glacier changes, and the accuracy can reach centimeter level.

## 4.3 Methodology

### *4.3.1 Landslide Recognizing Models*

#### 4.3.1.1 U-Net Model

The U-Net model is proposed to reduce the loss of information in feature maps and spatial locations during the pooling process of fully convolutional networks. In the model, the original image is first transformed into a high-dimensional feature vector by 4$\times$ downsampling with a fully connected neural network. Then the vector is restored to a feature map with the same resolution as the original image by 4$\times$ upsampling (Fig. 4.2). Each layer of the model has a long full connection, which ensures that the detailed features of each size can be preserved.

In the experiments, the model was trained for 200 epochs with a fixed learning rate of 0.001. A model trained with a learning rate of 0.01 was found less accurate

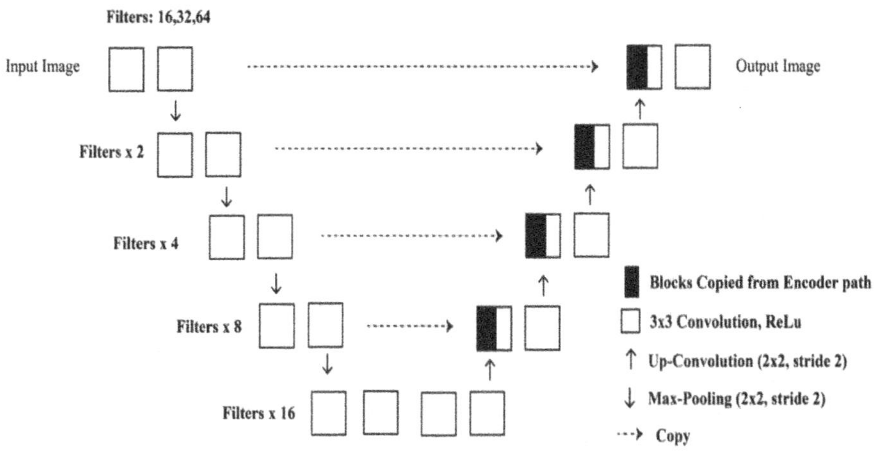


**Fig. 4.2** Structure diagram of U-Net

than the model trained with a learning rate of 0.001. Binary cross-entropy and Adam were used as loss function and optimization function, respectively. The model was only saved when the accuracy of the validation data increased to avoid overfitting. These models are built based on Keras and Tensorflow modules in Python.

#### 4.3.1.2 Random Forest

RF is a non-parametric machine learning classifier that has been proved to accurately distinguish spectrally complex classes (Belgiu & Drăguţ, 2016). An RF classifier is an ensemble classifier that grows multiple decision trees and trains them using bagging method, so that the trees can determine the probability of classes. RF can perform multi-criteria classification while being fast and insensitive to overfitting (Breiman, 2001).

RF model in this study was constructed by using the Tensorflow module in Python. To parameterize the RF classifier, based on out-of-bag accuracy, 200 trees were used and the square root of the number of layers was taken as the split criterion for each node. The sensitivity analysis method proposed by Stumpf and Kerle (2011) was considered, which tests different class ratios (β) to define the optimal ratio of training samples to find out the balance of user (overestimation) accuracy and producer (underestimation) accuracy. The parameter β refers to the ratio of landslide and non-landslide samples in the training set, and is changed iteratively to balance producer and consumer accuracy and get closer to the optimal value β. This study extracted landslide and non-landslide data in a large sample set and iterate 30 times for data in each year to test the optimal data ratio. The program started from an equal class distribution ($\beta = 1$), which defined the sample size of landslide (1000) and non-landslide (1000 × β). It is increased by 0.1 at each step β, until the five-fold of non-landslide rank distribution is reached (β = 3). For each iteration, the Stumpf and Kerle method were adjusted by using a consistent total sample size to avoid confounding sample size effects. In this way, 1000 training points of each year required for training were randomly selected from the large dataset, and 6000 verification points were sub-sampled from the dataset for each test year as the test set.

### 4.3.2 Data Sampling

In order to make a sample dataset that suitable for deep learning. The square masks with the size of multiple of 32 (32, 64, 128) were used to crop the image corresponding to the landslide catalog. The sampling process is shown in Fig. 4.3, and three sample datasets with different sizes were finally obtained as shown in Table 4.1. During RF training and sampling, the Moran index of landslides in the study area is relatively high, and the spatial correlation belongs to clustering, so points cannot be selected based on the Moran index. In addition, considering that the minimum number of pixels for a landslide is 4, a 120 m interval was used for equidistant regular sampling on the sampling images, and the number of samples for each year is shown in Table 4.1.

In Table 4.1, it can be observed RF sample pixels are about 8000–10,000, and non-landslide samples are about 20,000–30,000 for each year, and the distribution is relatively balanced. The number of U-Net samples is consistent every year, including 761 32 × 32 samples, 233 64 × 64 samples and 84 128 × 128 samples.

### 4.3.3 Model Performance Test

#### 4.3.3.1 The Influence of Data Dimension

Remote sensing images can be used to identify new landslides with drastic changes in spectral characteristics, which is suitable for interpretation of geological disasters after extreme events such as earthquakes and rainstorms. However, the spectral characteristics of the landslide surface are not unique. For example, some vegetation may be destroyed in the area of the new landslide, but some vegetation can be still preserved. Using spectral information may lead to wrong determinations of the landslide area. The old landslide is often covered with vegetations, and its spectrum features can be similar to that of agricultural and forestry plots. Therefore, methods based on spectral information have certain limitations. In addition, the performance remote sensing data is affected by the weather condition and environmental factors, so its ability to identify landslides will be restricted.

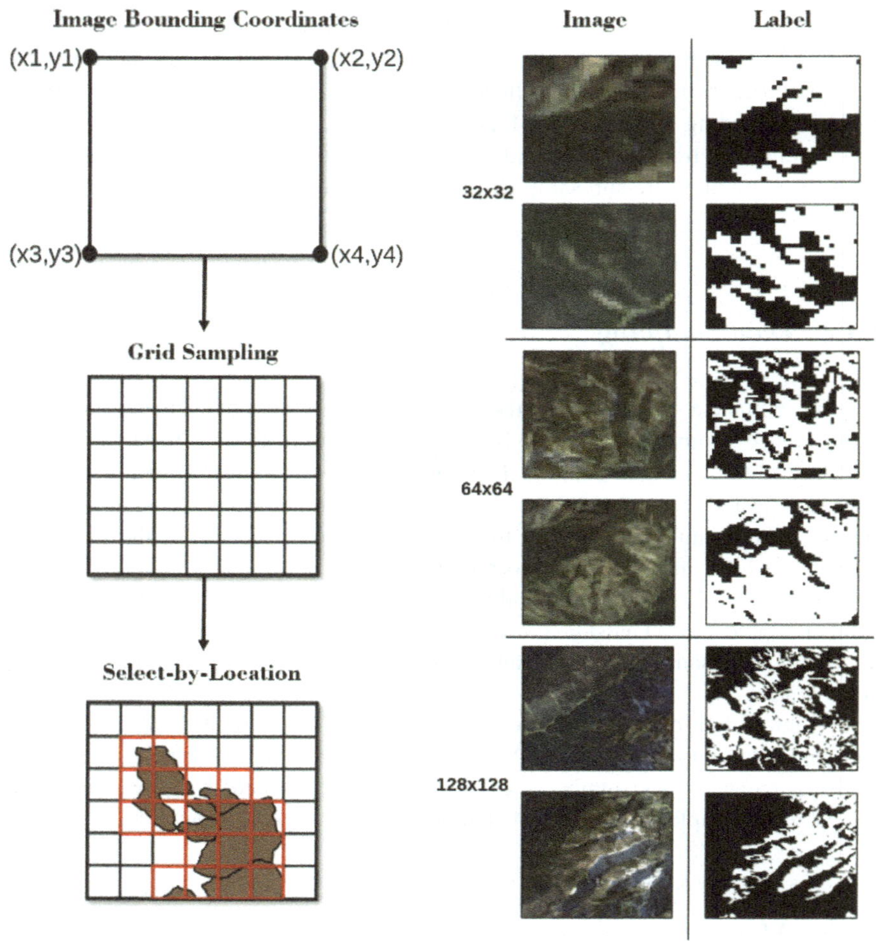


**Fig. 4.3** Example of the sampling process for a landslide scenario

**Table 4.1** Model training sample dataset

| Year | RF samples (pixels) | | U-Net samples (scenarios) | | |
|---|---|---|---|---|---|
| | Landslide | Non-landslide | 32 × 32 | 64 × 64 | 128 × 128 |
| 2008 | 10,131 | 27,415 | 761 | 233 | 84 |
| 2009 | 9520 | 23,807 | 761 | 233 | 84 |
| 2010 | 9173 | 24,086 | 761 | 233 | 84 |
| 2011 | 7067 | 30,416 | 761 | 233 | 84 |
| 2013 | 8861 | 24,438 | 761 | 233 | 84 |
| 2014 | 9316 | 24,164 | 761 | 233 | 84 |
| 2015 | 9974 | 27,559 | 761 | 233 | 84 |

In this case, the fusion of remote sensing image and DEM was tested to reveal whether including terrain information can help improve the accuracy of landslide identification. The performances of models with different input data (visible bands, all bands, all bands and DEM) were tested for each year. In the U-Net experiment, 70% of data in the annual dataset was randomly selected as the training data and the remaining 30% was used as the test data. After the RF model determined the optimal category ratio under different input channels, the pixels in the dataset were randomly selected in proportion to form the training set, and the remaining points were used as the verification set.

#### 4.3.3.2 Transfer Performance Evaluation

The samples from a single year were selected to construct the training set, and samples from the rest years for were used for verification to achieve the 10-year landslide mapping for the study area. In this way, the migration of models in different time dimension can be tested to reveal whether it can use only a few years of samples to carry out long-term series landslide mapping, thereby saving the sampling workload.

After conducting the abovementioned experiments, the results were taken to conduct annual dynamic detection of landslides from 2008 to 2018. The number of landslide pixels in each year was compared with the that in the catalog. A comparison chart of landslide area variation trend was drawn to test the accuracy of each model in describing the landslide activity change with the time.

### *4.3.4 Evaluation Metrics*

The performance of each model was evaluated by using accuracy rate, F1-score, recall, and precision. These metrics reflect whether a pixel is correctly classified as a landslide based on true positives (TP), false positives (FP) and false negatives (FN). FP represents a pixel is misclassified as landslide, while FN represents a pixel is misclassified as background.

Accuracy refers to the proportion of correctly detected samples to all samples. It is used to indicate the overall performance of the model, which is expressed by Eq. (4.1):

$$Accuracy = \frac{TP + TN}{TP + TN + FP + FN} \tag{4.1}$$

Recall calculates how many of the actual positives are TP. This metric is suitable for assessing the correlation of loss with FN, which is expressed by Eq. (4.2):

$$Recall = \frac{TP}{TP + FN} \tag{4.2}$$

Precision assesses how many pixels of the classified area are landslide. This is useful for assessing the loss of FP, which is expressed by Eq. (4.3):

$$Precision = \frac{TP}{TP + FP} \tag{4.3}$$

F1-score combines precision and recall to measure whether there is a balance between TP and FN, which is expressed by Eq. (4.4):

$$F1 - score = \frac{2 \cdot Precision \cdot Recall}{Precision + Recall} \tag{4.4}$$

## 4.4 Results

### *4.4.1 Data Channel Test*

In order to find a model that fits the characteristics of the study area and available dataset, we comprehensively tested the performance of the model with different inputs and sample sizes for 10 years. The test results are shown in Fig. 4.4. It can be observed that the landslide recognition accuracy of the U-Net model is significantly higher than that of the RF model in any case. The accuracy of the U-Net can be above 80% in most situations, reaching 90% in some scenarios (Fig. 4.4a). The RF is generally between 0.6 and 0.75 (Fig. 4.4b). It can be seen from Fig. 4.5 that the correct detection results account for a large proportion of the results of the U-Net model, which is significantly higher than the detection results of the RF model. Both figures prove that the performance of the U-Net model in landslide identification applications is significantly better than traditional pixel classification methods.

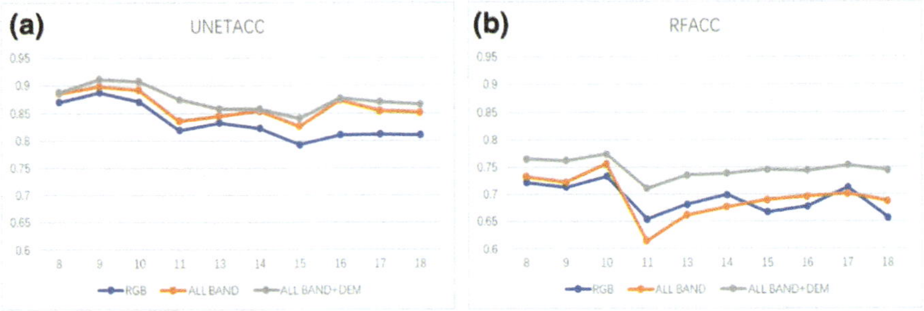


**Fig. 4.4** Channel and sample size tests of U-Net (**a**) and RF (**b**). The vertical axis represents the accuracy, and the horizontal axis represents the year. RGB represents the visible bands, ALL BAND represents all Landsat bands, and ALL BAND + DEM represents the fusion of all Landsat bands and DEM

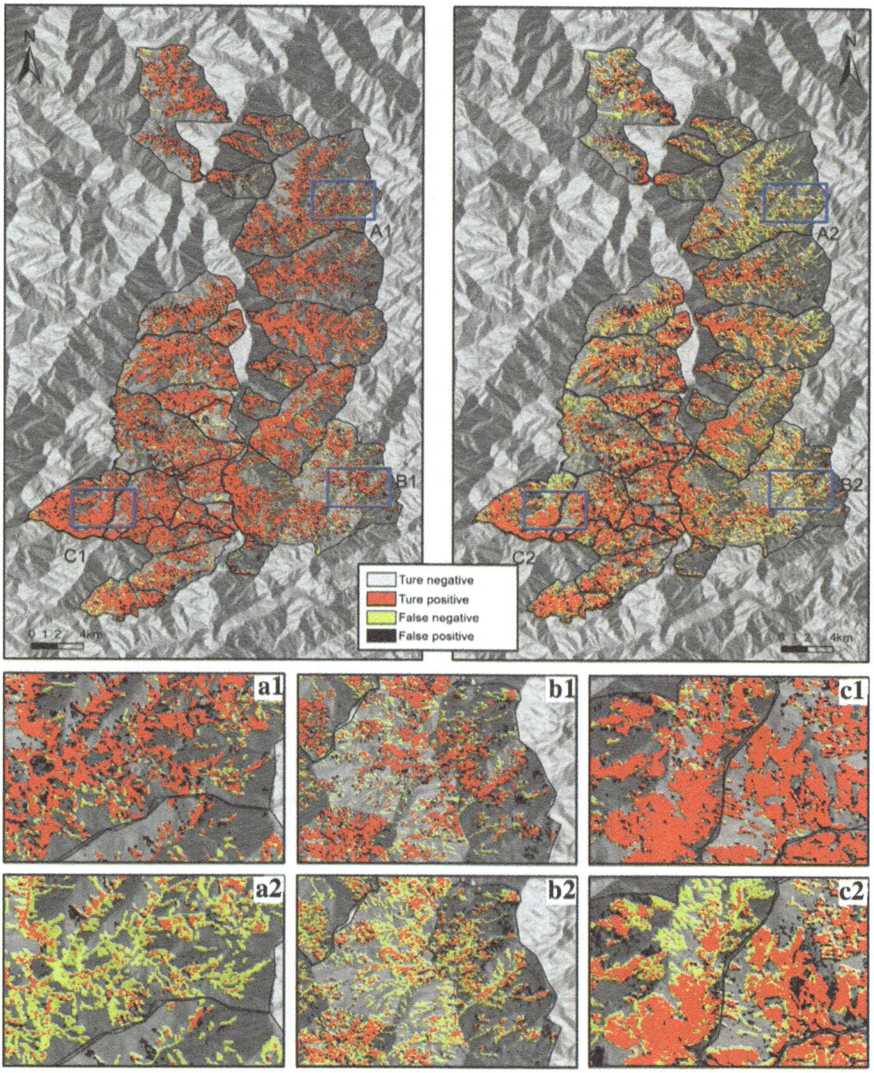


**Fig. 4.5** Landslide identification with U-Net model (top left) and random forest model (top right) (2015). **a1–c1** Part of the U-Net model recognition result; **a2–c2** part of the RF model recognition result

In addition, it can be observed from Fig. 4.4 that with more input features, the accuracy of the two methods for identifying landslides has been significantly improved. The difference in recognition accuracy between the model with only RGB and the model with fused remote sensing data and DEM is about 0.05 in most cases, even reaching 0.1 in some scenarios. Additional minable band information and terrain information provide more distinguishable features for landslide identification.

### 4.4.2 Temporal Transfer Capability of Models

The samples of a single year (2018) were selected to construct the training set, samples of the rest years were to construct the test set to analyze the time transfer capability of the model. The reason for choosing 2018 is that it is more in line with general practice to use the latest dataset to perform time transfer backwards. The transfer mapping accuracy of the two methods in each year is shown in Table 4.2, and the dynamic detection results of some years are shown in Fig. 4.6.

It can be seen from Table 4.2 that the evaluation metrics of U-Net model transfer learning in most years are close to 80%, which is significantly higher than that of RF model. Especially in years between 2014 and 2018, most metrics are close to or exceed 90%, indicating high performances of dynamic monitoring. This shows that the time transfer capability of U-Net is acceptable to achieve the long-term landslide dynamic detection with samples in limited years. This provides a reliable technical means for the dynamic detection of historical landslides at a lower cost. The metrics of RF model are generally lower than 80%. And the accuracy is lower in the early years, so it is unable to meet the generally required 80% drawing accuracy requirements. Studies have shown that due to the long-term persistence of landslides, there is no significant changes of most of the landslides in different years, and only a few landslides are newly added or disappeared. Therefore, the accuracy of landslide recognition does not change too much during the time transfer learning for long time series. For the higher-precision U-Net model, the accuracy of each year can be always acceptable when carrying out time series transfer learning.

Figures 4.6 and 4.7 indicate the agreement between the model detection results and the reference data. In Fig. 4.6, the correct detection (TP) of the U-Net model is significantly higher than that of the RF model in each year. In Fig. 4.7, the detection curve of the U-Net model is basically consistent with the curve of the reference catalog data, which indicates a good agreement. In the RF model, a large number

**Table 4.2** Migration mapping accuracy of the two models

| Year | RF | | | U-Net | | |
|---|---|---|---|---|---|---|
| | Accuracy % | Precision % | Recall % | Accuracy % | Precision % | Recall % |
| 2008 | 73.54 | 69.32 | 70.20 | 86.59 | 85.41 | 83.83 |
| 2009 | 74.18 | 70.27 | 68.35 | 84.42 | 78.90 | 85.08 |
| 2010 | 73.36 | 68.66 | 70.34 | 82.75 | 74.55 | 85.31 |
| 2011 | 71.12 | 68.55 | 65.42 | 81.77 | 71.67 | 85.02 |
| 2013 | 73.51 | 66.23 | 68.20 | 84.50 | 79.88 | 84.25 |
| 2014 | 73.83 | 67.75 | 69.92 | 85.37 | 76.39 | 89.13 |
| 2015 | 74.51 | 69.02 | 66.73 | 93.85 | 92.59 | 93.81 |
| 2016 | 74.33 | 69.70 | 67.83 | 90.10 | 84.25 | 92.61 |
| 2017 | 75.31 | 70.19 | 71.30 | 95.42 | 95.64 | 94.07 |
| 2018 | 79.41 | 78.77 | 75.66 | 97.63 | 97.51 | 97.12 |

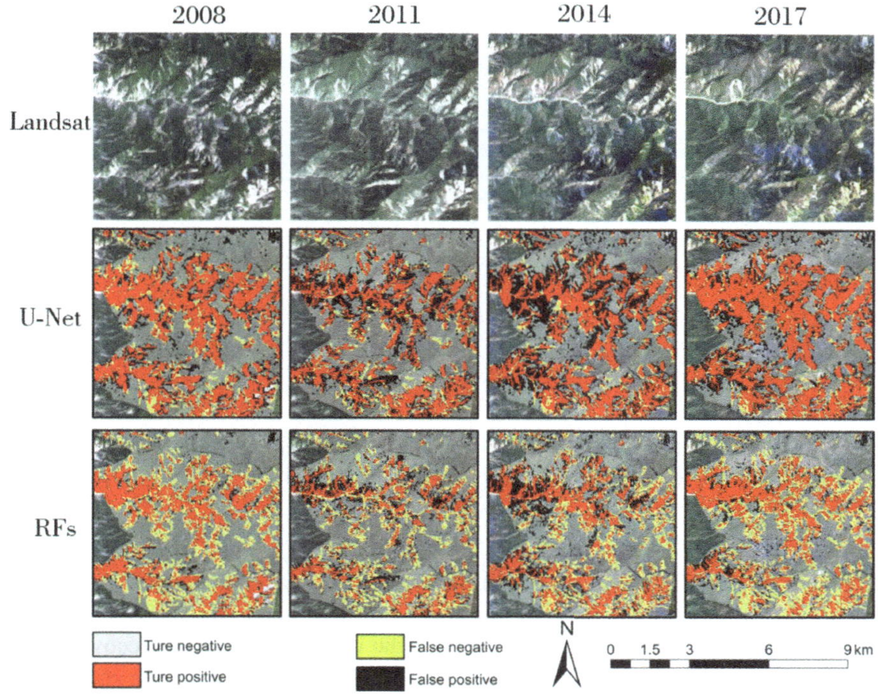


**Fig. 4.6** Detection of transfer learning dynamics for some years for both models

of false detections (FN and FP) can be seen in Fig. 4.6, and the difference in the detection results in Fig. 4.7 is significant.

### *4.4.3 Spatio-Temporal Dynamic Detection of Landslides*

The previous experiments have proved the superior performance of the U-Net model in the dynamic detection of time series landslides. On this basis, the sample set in 2018 was used to carry out the dynamic detection of landslides in 2019–2021 from remote sensing images and DEM, and the results are shown in Fig. 4.8. It can be observed that the number of landslide pixels in 2019 decreased by about 10% compared to 2018, that in 2020 decreased by about 14% comparing to 2019, that in 2021 decreased by about 2% comparing to 2020. Overall, the analysis of landslide changes from 2008 to 2021 indicates that most of the landslides in the study area were caused by earthquakes. Over time, the earthquake-stricken area gradually became stabilized, so the number of landslides gradually decreased after the earthquakes, and the surface vegetation recovered (Zhong et al., 2021). The Ya'an earthquake in 2013 led to some serious landslides. Since then, due to the slow accumulation of soil

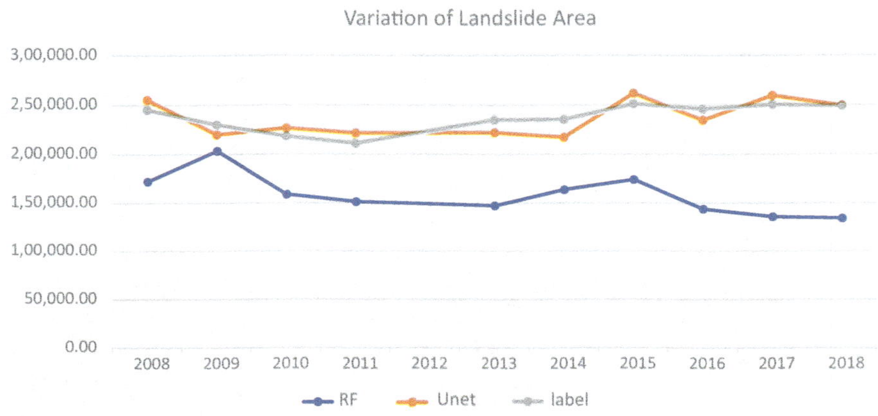


**Fig. 4.7** General trend of landslide dynamic detection. The vertical axis represents the number of pixels of the landslide, the horizontal axis represents the year, and the 'label' represents the reference landslide catalog data

that were suitable for plant growth on the landslide surface, the landslide area was stable and the vegetation recovered slowly. After more than 10 years, the soil area on the surface of the landslide was stable and the vegetation coverage recovered. As no more earthquake occurred, the area of the landslide can keep getting smaller.

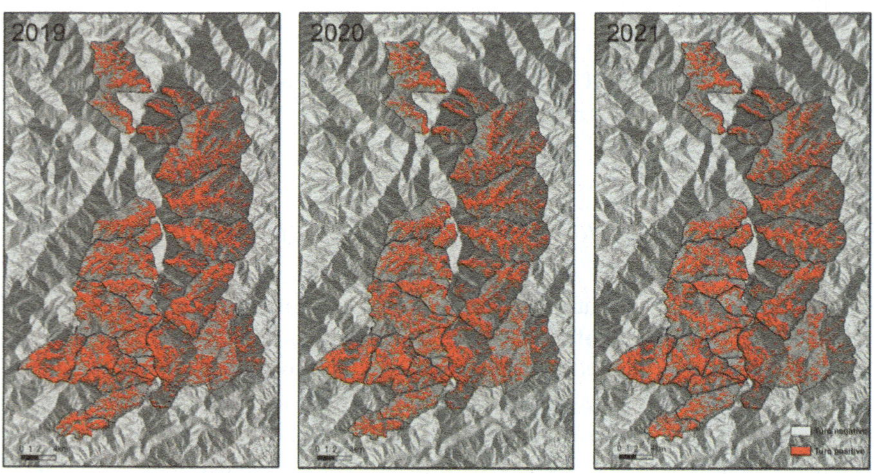


**Fig. 4.8** Landslide detection in 2019–2021. The red pixels represent the recognition results

## 4.5 Discussion

In recent years, the impact of human activities (carbon dioxide emissions, major projects, etc.) on the ecological environment is getting significant, leading to the global climate change and frequent weather extremes (Chou et al., 2013). Studies have shown that the increase of the intensity and frequency of extreme precipitation caused by climate change will lead to more mountain landslide events (Chang et al., 2014; Wei et al., 2018). In addition to the shallow landslides directly induced by extreme precipitation, global warming can also indirectly affect landslide susceptibility by changing landslide related geological environmental factors (such as permafrost, vegetation cover, runoff, etc.) (Zhao et al., 2019). "Third National Assessment Report on Climate Change" released by the Ministry of Science and Technology of China indicates that in the following 50 years, the average annual precipitation in China will keep the increasing trend. Weather extremes such as high temperature, typhoon, and storm surge will become more frequent, which may result in floodings, landslides, debris flows, and exacerbated the drought and land degradation. In addition, the rise of air temperature on the Qinghai-Tibet Plateau leads to accelerated degradation of glaciers and permafrost, which thereby causes higher slope instability and the higher possibility of mountain disasters (Cui et al., 2019).

Landslide investigation is necessary for the understanding of the landslide hazards distribution in the region and disaster prevention and mitigation. However, there are great uncertainties regarding the location, time, scale and causes of the landslides. And the complexity of environmental factors such as geological structure and topography also brings great difficulties to the investigation of landslide potential areas (Xu et al., 2019). How to efficiently identify landslide potential areas and to support the formulation of disaster prevention and early warning strategies is still a key issue in the current geological disaster prevention and control work. The development of remote sensing technology allows the multi-scale and continuous observation of the earth surface to be achieved (Zhao et al., 2020). Time-series remote sensing images record surface variations and trajectories in detail, which can continuously monitor the changes in land use/cover, vegetation, topography, rock-soil properties and other disaster-forming factors for landslides in long-term series, and detect the occurrence of earthquakes, precipitation extremes and indicate their influence evolution process (Liu et al., 2018; Xing & Niu, 2019).

Landslides are widespread, but are only mapped in detail in limited scales and within a short period of time. Currently the information in landslide database is not adequate. For example, whether the old landslide areas have been recovered or restored have not been updated. In the Wenchuan earthquake area, the time series dynamic changes of landslides from 2008 to 2021 were studied in this research. In the future, the variations will still be monitored to explore the impact of earthquakes on regional stability. This study is of great significance for revealing the long-term impact of major events and stability of the region. In this study, it is found that the number and area of landslide activity gradually decreased over time, indicating that the impact of earthquakes on the region is weakening.

This study has some limitations. Firstly, this study did not perform the landslide identification for the pre-earthquake years. This is because the landslides during this period were not induced by earthquakes, which is not helpful for analyzing the impact of earthquakes on the long-term stability of the region. In addition, the area of landslides before the earthquake is relatively small, with few amounts, and sparse distribution, so it is difficult to use Landsat image and deep learning for the recognition. Also, limited by the missing pixels of the Landsat-7 images, the landslide dynamic detection study in 2012 was not performed in this study. Higher-resolution Sentinel-2 MSI time series images were not used in this study as they cannot fully cover the time series from 2008 to 2021. However, this data can significantly improve the precision of landslide identification, and it can be tried to be applied in other time series change detection.

Despite these limitations, it is still recommended to use the cost-effective data and methods for long-term landslide dynamic monitoring like this. Because it can help to understand the spatiotemporal activity of landslides, regional stability and vegetation recovery, which can provide scientific data for regional geological disaster prevention, land use management, economic construction and ecological protection.

## 4.6 Summary

In this study, long-term Landsat images and DEM were integrated to conduct long-term dynamic monitoring of landslides in the severely affected area of the Wenchuan Earthquake. The study found that the U-Net model can achieve much better detection accuracy than RF model in all scenarios. Integration multi-source data helps the model to well understand the landslides and significantly improves the accuracy of landslide identification. In addition, the U-Net model also shows a fairly high accuracy in non-training years, indicating the good time invariance in landslide identification. On this basis, variation of landslides from 2008 to 2021 were studied. In the future, the landslide variation will still be monitored in the study area to explore the impact of earthquakes on regional stability. This study contributes to the development of remote sensing time series analysis technology and landslide remote sensing identification method, and also help to promote the extensive development of historical landslide cataloging and long-term landslide activity analysis, which will better support the major engineering construction and mountainous land development and management.

**Codes**

**[Google Earth Engine] Random Forest Classification**

```
//Page1 Data acquisition and train set ratio
var roi = ee.Geometry.MultiPolygon(
      [[[103.35480248300236, 31.430527847244427],
        [103.35480248300236, 30.966530530644683],
```

```
        [103.67477928964298, 30.966530530644683],
        [103.67477928964298, 31.430527847244427]]], null, false);
Map.centerObject(roi, 10);
Map.addLayer(roi, {color: "red"}, "roi");
//Landsat FMASK
function rmCloud(image) {
 var cloudShadowBitMask = (1 << 5);
 var cloudsBitMask = (1 << 3);
 var qa = image.select("QA_PIXEL");
 var mask = qa.bitwiseAnd(cloudShadowBitMask).eq(0)
         .and(qa.bitwiseAnd(cloudsBitMask).eq(0));
 return image.updateMask(mask);
}
//scaleImage
function scaleImage(image) {
 var time_start = image.get("system:time_start");
 image = image.multiply(0.000010);
 image = image.set("system:time_start", time_start);
 return image;
}
var ColSP08 = ee.ImageCollection("LANDSAT/LT05/C02/T1_L2")
          .filterBounds(roi)
          .filterDate("2008-5-1", "2008-12-31")
          .map(rmCloud)
          .map(scaleImage)
//DEM
var srtm = ee.Image("USGS/SRTMGL1_003");
var dem = ee.Algorithms.Terrain(srtm);
var elevation = dem.select("elevation");
var slope = dem.select("slope");
var aspect = dem.select("aspect");
var sinImage = aspect.divide(180).multiply(Math.PI).sin();
var Landsat_DEM= [
 "SR_B1", "SR_B2", "SR_B3", "SR_B4", "SR_B5","SR_B7",
  "SLOPE", "ELEVATION","ASPECT"
];
var RGB=[
  "SR_B2", "SR_B3", "SR_B4"
]
var Landsat= [
 "SR_B1", "SR_B2", "SR_B3", "SR_B4", "SR_B5","SR_B7"
]
var Image08 = ColSP08.median()
           .clip(roi).addBands(elevation.rename("ELEVATION"))
           .addBands(slope.rename("SLOPE"))
           .addBands(sinImage.rename("ASPECT"))
           .select(Landsat_DEM);
var rgbVisParam = {
 min: 0,
 max: 0.3,
 bands: ["SR_B3", "SR_B2", "SR_B1"]}
Map.addLayer(Image08, rgbVisParam, "image2008");
//show datasat
```

```
//var Region = ee.FeatureCollection("projects/ee-hzk/assets/WC/
WC2008")
var Region = ee.FeatureCollection('projects/ee-hzk/assets/WC/
WC2008')
var styling={color:'#fafafa',fillColor:'00000000',width:2}
var fcVis = Region.style(styling)
Map.addLayer(Region,{},'STD')
//Split training and test dataset
var landslide8= Image08.sample({
  region:ee.FeatureCollection('projects/ee-hzk/assets/WC/WC2008
'),
 geometries: true,
 scale:120
});
var pionts8= Image08.sample({
 region:ee.FeatureCollection(roi),
 geometries: true,
 scale: 200
});
var Region8 = ee.FeatureCollection('projects/ee-hzk/assets/WC/
WC2008')
var continentalUS8 = ee.Geometry(Region8);
var r8 = ee.Filter.bounds(Region8)
var nr8 =r8.not()
var no_filtered8= pionts8.filter(nr8)
var landslide8=landslide8.map(
   function(feat){
   feat = feat.set('lc_numeric',1);
   return feat
  }
)
var no_landslide8=no_filtered8.map(
   function(feat){
   feat = feat.set('lc_numeric',0);
   return feat
  }
)
var landslidesize = landslide8.size()
var no_landslidesize = no_landslide8.size();
var no_id = landslide8.merge(no_landslide8);
var wc2008data = no_id.randomColumn('id',5);
print("landslide_size", landslidesize)
print("no_landslide_size", no_landslidesize )
//ratio
var Epoch = 1
var i = 0
for (i = 0; i <= Epoch; i++) {
var x = 1.0 + (i*0.1)
var samplenumber= ee.Number(1000.0)
var sample_landslide =
wc2008data.filterMetadata('lc_numeric','equals',1).    random-
Column('random',i);
var   sample_no_landslide   =   wc2008data.filterMetadata('lc_
numeric','equals',0).randomColumn('random',i);
```

```
var landslidetrain_sample_ratio = samplenumber.
divide(landslidesize)
var nolandslidetrain_sample_ratio = samplenumber.multiply(x).
divide(no_landslidesize)
var sample_training_landslide =sample_landslide.
filter(ee.Filter.lte('random', landslidetrain_sample_ratio));
var sample_training_nolandslide =sample_no_
landslide.filter(ee.Filter.lte('random', nolandslidetrain_
sample_ratio));
var ratio_train_data = sample_training_landslide.merge(sample_
training_nolandslide)
print (ratio_train_data.size(),"trainsize")
var joinFilter = ee.Filter.equals({leftField:'id', right-
Field:'id'}); //defined a connection with equals
var sample_v = ee.Join.inverted().apply(wc2008data, ratio_train_
data, joinFilter).randomColumn('randomv',i); // Jioned FeatureC-
ollection
var sample_v_size = sample_v.size()
var samplenumberV= ee.Number(6000)
var sample_v_ratio = samplenumberV.divide(sample_v_size)
var sample_validate = sample_v.filter(ee.Filter.lte('randomv',
landslidetrain_sample_ratio));
print (sample_validate.size(),"testsize")
// Make a Random Forest classifier and train it.
var classifier = ee.Classifier.smileRandomForest(200)
   .train({
    features: ratio_train_data,
    classProperty: 'lc_numeric',
   inputProperties: [ "SR_B1", "SR_B2", "SR_B3", "SR_B4", "SR_B5",
"SR_B7",
 "SLOPE", "ELEVATION","ASPECT" ]
   });
//Train data confusion matrix
var trainAccuracy = classifier.confusionMatrix();
var kappa = trainAccuracy.kappa()
var matrix= trainAccuracy.array()
print(kappa,'kappa')
print(matrix,'matrix')
print(trainAccuracy.accuracy(),'trainAccuracy')
//validation
var validated = sample_validate.classify(classifier);
//Test confusion matrix
var testAccuracy = validated.errorMatrix("lc_numeric", "classifi-
cation");
var matrixt= testAccuracy.array()
print(x ,'ratio')
print(matrixt,'matrixt')
print(testAccuracy.accuracy(),'testAccuracy')
 }
Export.table.toDrive(wc2008data,
'wc2008data',
'WC_data',
'wc2008data',
'CSV')
```

```
//Page2 Train set construction and modelling for image recognition
var roi = ee.Geometry.MultiPolygon(
      [[[103.35480248300236, 31.430527847244427],
        [103.35480248300236, 30.966530530644683],
        [103.67477928964298, 30.966530530644683],
        [103.67477928964298, 31.430527847244427]]], null, false);
Map.centerObject(roi, 10);
Map.addLayer(roi, {color: "red"}, "roi");
//Landsat FMASK
function rmCloud(image) {
 var cloudShadowBitMask = (1 << 5);
 var cloudsBitMask = (1 << 3);
 var qa = image.select("QA_PIXEL");
 var mask = qa.bitwiseAnd(cloudShadowBitMask).eq(0)
           .and(qa.bitwiseAnd(cloudsBitMask).eq(0));
 return image.updateMask(mask);
}
//scaleImage
function scaleImage(image) {
 var time_start = image.get("system:time_start");
 image = image.multiply(0.000010);
 image = image.set("system:time_start", time_start);
 return image;
}
var ColSP08 = ee.ImageCollection("LANDSAT/LT05/C02/T1_L2")
          .filterBounds(roi)
          .filterDate("2008-5-1", "2008-12-31")
          .map(rmCloud)
          .map(scaleImage)
//DEM
var srtm = ee.Image("USGS/SRTMGL1_003");
var dem = ee.Algorithms.Terrain(srtm);
var elevation = dem.select("elevation");
var slope = dem.select("slope");
var aspect = dem.select("aspect");
var sinImage = aspect.divide(180).multiply(Math.PI).sin();
var Landsat_DEM= [
 "SR_B1", "SR_B2", "SR_B3", "SR_B4", "SR_B5","SR_B7",
  "SLOPE", "ELEVATION","ASPECT"
];
var RGB=[
  "SR_B2", "SR_B3", "SR_B4"
]
var Landsat= [
 "SR_B1", "SR_B2", "SR_B3", "SR_B4", "SR_B5","SR_B7"
]
var Image08 = ColSP08.median()
              .clip(roi).addBands(elevation.rename("ELEVATION"))
              .addBands(slope.rename("SLOPE"))
              .addBands(sinImage.rename("ASPECT"))
              .select(Landsat_DEM);

var rgbVisParam = {
 min: 0,
```

```
 max: 0.3,
 bands: ["SR_B3", "SR_B2", "SR_B1"] }
Map.addLayer(Image08, rgbVisParam, "image2008");
//show datasat
//var Region = ee.FeatureCollection("projects/ee-hzk/assets/WC/
WC2008")
var Region = ee.FeatureCollection('projects/ee-hzk/assets/WC/
WC2008')
var styling={color:'#fafafa',fillColor:'00000000',width:2}
var fcVis = Region.style(styling)
Map.addLayer(Region,{},'STD')
//Split training and testing datasets
var landslide8= Image08.sample({
  region:ee.FeatureCollection('projects/ee-hzk/assets/WC/WC2008
'),
 geometries: true,
 scale:120
});
var pionts8= Image08.sample({
 region:ee.FeatureCollection(roi),
 geometries: true,
 scale: 200
});
var Region8 = ee.FeatureCollection('projects/ee-hzk/assets/WC/
WC2008')
var continentalUS8 = ee.Geometry(Region8);
var r8 = ee.Filter.bounds(Region8)
var nr8 =r8.not()
var no_filtered8= pionts8.filter(nr8)
var landslide8=landslide8.map(
   function(feat){
   feat = feat.set('lc_numeric',1);
   return feat
  }
)
var no_landslide8=no_filtered8.map(
   function(feat){
   feat = feat.set('lc_numeric',0);
   return feat
  }
)
var landslidesize = landslide8.size()
var no_landslidesize = no_landslide8.size();
var no_id = landslide8.merge(no_landslide8);
var wc2008data = no_id.randomColumn('id',5);
print("landslide_size", landslidesize)
print("no_landslide_size", no_landslidesize )


//R is the best ratio of file1, manually filled.
var R=ee.Number(2.2)
var no_landslide8 = no_landslide8.randomColumn('randomratio')
var nolandslide_ratio = landslidesize.multiply(R).divide(no_
landslidesize)
```

```
var nolandlides = no_landslide8.filter(ee.Filter.lte('randomratio',
nolandslide_ratio));
var wc8dataset = nolandlides.merge(landslide8).randomColumn('slice',
2);
var training8 =wc8dataset.filter(ee.Filter.lte('slice', 0.7));
var joinFilter = ee.Filter.equals({leftField:'id', right-
Field:'id'}); //defined a connection with equals
var validate8 = ee.Join.inverted().apply(wc2008data, training8,
joinFilter);
Export.table.toAsset(training8,
'training8',
'WC_data',
'CSV')
Export.table.toAsset(validate8,
'validate8',
'WC_data',
'CSV')
// Make a Random Forest classifier and train it.
var classifier = ee.Classifier.smileRandomForest(200)
   .train({
    features:training8,
    classProperty: 'lc_numeric',
   inputProperties: [ "SR_B1", "SR_B2", "SR_B3", "SR_B4", "SR_B5",
"SR_B7",
"SLOPE", "ELEVATION","ASPECT" ]
   });
//TD confusion matrix
var trainAccuracy = classifier.confusionMatrix();
var kappa = trainAccuracy.kappa()
var matrix= trainAccuracy.array()
print(kappa,'kappa')
print(matrix,'matrix')
print(trainAccuracy.accuracy(),'trainAccuracy')
//Validation
var validated = validate8.classify(classifier);
var testAccuracy = validated.errorMatrix("lc_numeric", "classifi-
cation");
var matrixt= testAccuracy.array()
print(matrixt,'matrixt')
print(testAccuracy.accuracy(),'testAccuracy')
print('AconsumersAccuracy',testAccuracy.consumersAccuracy());
print('AaproducersAccuracy',testAccuracy.producersAccuracy());


// Use training data training to obtain classification results
var classified =Image08.classify(classifier);
var igbpPalette = [
  'aec3d4', // lan
  '111149', // no
];
Map.addLayer(classified.clip(roi),
         {palette: igbpPalette, min: 0.5, max: 1},
         'classification');
// Map.addLayer(STD)
```

```
var Region = ee.FeatureCollection("projects/ee-hzk/assets/WC/
WC2008")
var styling={color:'#fafafa',fillColor:'00000000',width:2}
var fcVis = Region.style(styling)// modify to own feature
Map.addLayer(Region,{},'STD')
Export.image.toAsset({
 image: classified,
 scale: 30,
 maxPixels: 1e13,
 region: roi,
 description: '2008RFD',
 assetId: 'WC2008RFD'
})
```

**[Python3.9] U-Net Classification Data Preprocessing**

```
from google.colab import auth
auth.authenticate_user()
# Import, authenticate and initialize the Earth Engine library.
import ee
ee.Authenticate()
ee.Initialize()
# Folium setup.
import folium
print(folium.__version__)
# Use Landsat 5 surface reflectance data.
roi1 = ee.Geometry.Polygon(
      [[[103.35480248300236, 31.430527847244427],
      [103.35480248300236, 30.966530530644683],
      [103.67477928964298, 30.966530530644683],
      [103.67477928964298, 31.430527847244427]]], None, False) )
#recognising scale
landsat5src01 = ee.ImageCollection("LANDSAT/LT05/C01/T1_SR")
opticalBands = ['B1', 'B2', 'B3', 'B4', 'B5', 'B6', 'B7']
roi = ee.FeatureCollection('projects/ee-hzk/assets/FCNWC/roi')
# Cloud masking function.
def maskL5sr(image):
 cloudShadowBitMask = ee.Number(2).pow(2).int()
 cloudsBitMask = ee.Number(2).pow(5).int()
 qa = image.select('pixel_qa')
 mask1 = qa.bitwiseAnd(cloudShadowBitMask).eq(0).And(
  qa.bitwiseAnd(cloudsBitMask).eq(0))
 mask2 = image.mask().reduce('min')
 mask3 = image.select(opticalBands).gt(0).And(
      image.select(opticalBands).lt(10000)).reduce('min')
 mask = (mask2).And(mask3)
 return image.select(opticalBands).divide(10000).updateMask(mask)
# The image input data is a cloud-masked median composite.
srtm = ee.Image("USGS/SRTMGL1_003");
dem = ee.Algorithms.Terrain(srtm);
elevation = dem.select("elevation").divide(10000);
slope = dem.select("slope").divide(100);
aspect = dem.select("aspect");
```

```
sinImage = aspect.divide(180).sin().float();
#image_collection = (landsat5src01.filterDate('2010-3-1', '2010-
4-1')    .merge(landsat5src01.filterDate('2010-5-1',    '2010-7-
1'))  .merge(landsat5src01.filterDate('2011-5-1',  '2011-8-1')))
.map(maskL5sr).min()
#Multi-temporal data fusion
image_collection  =  landsat5src01.filterDate('2008-5-1',  '2009-
12-31').map(maskL5sr).min()
image = image_collection.where(image_collection.gte(0.30),0.3).
where(image_collection.lte(0),0).select('B7','B5','B4',"B3",
"B2","B1").rename('B7','B6','B5',"B4","B3","B2").addBands
(elevation .rename("ELEVATION")).addBands(slope.rename("SLOPE"))
.addBands (sinImage.rename("ASPECT"))
Null_label    =    landsat5src01.filterDate('2010-1-1','2010-12-
1').median().select("B4").uint8().rename("label").multiply(0).
clip(roi1)
labelshp  =  ee.FeatureCollection("projects/ee-hzk/assets/FCNWC/
WC8")
image_reduced= labelshp.reduceToImage(['N'],ee.Reducer.first()).
divide(3).unmask(0).float()
label= Null_label.add(image_reduced).unmask(0)
image = image.addBands(label)
# Use folium to visualize the imagery.
mapid  =  image.getMapId({'bands':  ['B4',  'B3',  'B2'],  'min':  0,
'max': 0.3})
map = folium.Map(location=[31.5,103.03 ])
folium.TileLayer(
   tiles=mapid['tile_fetcher'].url_format,
   attr='Map Data &copy; <a href="https://earthengine.google.com/
">Google Earth Engine</a>',
   overlay=True,
   name='median composite',
  ).add_to(map)
mapid = image.getMapId({'bands': ['label']})
folium.TileLayer(
   tiles=mapid['tile_fetcher'].url_format,
   attr='Map Data &copy; <a href="https://earthengine.google.com/
">Google Earth Engine</a>',
   overlay=True,
   name='label',
  ).add_to(map)
map.add_child(folium.LayerControl())
map
pathch_sample=image
KERNEL_SIZE = 128
list = ee.List.repeat(1, KERNEL_SIZE)
lists = ee.List.repeat(list, KERNEL_SIZE)
kernel = ee.Kernel.fixed(KERNEL_SIZE, KERNEL_SIZE, lists)
arrays = pathch_sample.neighborhoodToArray(kernel) #Image
print(arrays.getInfo())
Bands  =  ['B2',  'B3',  'B4',  'B5',  'B6',  'B7',"ELEVATION","SLOPE",
"ASPECT",'label']
RGB = ['B2', 'B3', 'B4']
DEM = ["ELEVATION","SLOPE","ASPECT"]
```

```
label = ['label']
#Output label raster
Bands = ['B2', 'B3', 'B4', 'B5', 'B6', 'B7',"ELEVATION","SLOPE",
"ASPECT"]
#Bands = ['SR_B2', 'SR_B3', 'SR_B4', 'SR_B5', 'SR_B6', 'SR_
B7',"ELEVATION","SLOPE","ASPECT"]
def doExport(out_image_base,region):
  """Run the image export task. Block until complete.
  """
  task = ee.batch.Export.image.toDrive(
   image = image.select(label),
   description = out_image_base,
   folder= 'label08',
   fileNamePrefix = out_image_base,
   region = region.getInfo()['coordinates'],
   scale = 30,
   fileFormat = 'GeoTIFF',
   maxPixels = 1e10,
  )
  task.start()
# Base file name to use for TFRecord files and assets.
out_image_base = 'label08'
# roi
cliproi = ee.Geometry.Polygon(
        [[[103.35480248300236, 31.430527847244427],
        [103.35480248300236, 30.966530530644683],
        [103.67477928964298, 30.966530530644683],
        [103.67477928964298, 31.430527847244427]]], None, False) #
recognising scale
# Run the export.
doExport(out_image_base,cliproi)
#Output train set
sample_grid = ee.FeatureCollection('projects/ee-hzk/assets/
FCNWC/P128')#FeatureCollection
n=40
sample_num=sample_grid.size().getInfo()
for i in range(0,sample_num-1,n):
    sample_grid_list=sample_grid.toList(n,i) #number of output,
start number
   pathch_sample=ee.FeatureCollection(sample_grid_list)
   sample = arrays.sampleRegions(
      collection = pathch_sample,
        scale = 30,
        tileScale=4)
   desc =str(i)
   print(desc)
   task = ee.batch.Export.table.toDrive(
      collection = sample,
      description = desc,
      folder= 'train2008',
      fileNamePrefix = desc,
      fileFormat = 'TFRecord',
      selectors = Bands, #[R,B,G,Label]
        )
```

```
    task.start()
#Recognition image export
# If the recognized image is 128*128, it is actually cropped to
(128+64) * (128+64)
def doExport(out_image_base,kernel_buffer,region):
  """Run the image export task. Block until complete.
  """
  task = ee.batch.Export.image.toDrive(
    image = image.select(RGB),
    description =out_image_base,
    folder= 'WC2008D',
    fileNamePrefix =out_image_base,
    region = region.getInfo()['coordinates'],
    scale = 30,
    fileFormat = 'TFRecord',
    maxPixels = 1e10,
    formatOptions = {
      'patchDimensions': [128, 128],
      'kernelSize': kernel_buffer,
      'compressed': True,
      'maxFileSize': 104857600
    }
  )
  task.start()
# Output assets folder: YOUR FOLDER
user_folder = 'projects/ee-hzk/assets/FCNWC' # INSERT YOUR FOLDER
HERE.
# Base file name to use for TFRecord files and assets.
wc_image_base = 'wc2008d'
# Half this will extend on the sides of each patch.
wc_kernel_buffer = [64,64]
# roi
cut = ee.Geometry.Polygon(
        [[[103.35480248300236, 31.430527847244427],
        [103.35480248300236, 30.966530530644683],
        [103.67477928964298, 30.966530530644683],
        [103.67477928964298, 31.430527847244427]]], None, False)
# recognising scale
# Run the export.
doExport(wc_image_base, wc_kernel_buffer,cut)
```

## [Python3.9] U-Net Classification Inference

```
import tensorflow as tf
import numpy as np
import os
from osgeo import gdal,osr
MODEL_DIR = r'model_path'
m = tf.keras.models.load_model(MODEL_DIR)
m.summary()
kernel_buffer = [64, 64]
KERNEL_SIZE = 128
KERNEL_SHAPE = [KERNEL_SIZE, KERNEL_SIZE]
```

```
Bands = ['B2', 'B3', 'B4', 'B5', 'B6', 'B7',"ELEVATION","SLOPE",
"ASPECT"] #landsat+dem
#Bands = ['B2', 'B3', 'B4']#RGB
#Bands = ['B2', 'B3', 'B4', 'B5', 'B6', 'B7']#landsat
"""Perform inference on exported imagery, upload to Earth
Engine."""
def doPrediction(out_image_base, base_dir):
    """Perform inference on exported imagery, upload to Earth
Engine."""
# Get all tif files to predict
  print('Looking for TFRecord files...')
  # Get a list of all the files in the output bucket.
  filesList = os.listdir(base_dir)
  # Get only the files generated by the image export.
  exportFilesList = [base_dir+s for s in filesList if out_image_
base in s]
  # Get the list of image files and the JSON mixer file.
  imageFilesList = []
  jsonFile = None
  for f in exportFilesList:
    if f.endswith('.tfrecord.gz'):
      imageFilesList.append(f)
    elif f.endswith('.json'):
      jsonFile = f
  # Make sure the files are in the right order.
  imageFilesList.sort()
  from pprint import pprint
  pprint(imageFilesList)
  print(jsonFile)
# Parse json file
  import json
  # Load the contents of the mixer file to a JSON object.
  # jsonText = !gsutil cat {jsonFile}

  # Get a single string w/ newlines from the IPython.
utils.text.SList
  mixer = json.load(open(jsonFile))
  pprint(mixer)
  patches = mixer['totalPatches']
# Get set up for prediction.
  buffered_shape = [
    KERNEL_SHAPE[0] + kernel_buffer[0],
    KERNEL_SHAPE[1] + kernel_buffer[1]]
  imageColumns = [
    tf.io.FixedLenFeature(shape=buffered_shape,dtype=tf.
float32)
    for k in Bands
  ]
  imageFeaturesDict = dict(zip(Bands, imageColumns))
# Data Analysis
  def parse_image(example_proto):
     return tf.io.parse_single_example(example_proto, imageFea-
turesDict)
  def toTupleImage(inputs):
```

```
      inputsList = [inputs.get(key) for key in Bands]
      stacked = tf.stack(inputsList, axis=0)
      stacked = tf.transpose(stacked, [1, 2, 0])
      return stacked
   # Create a dataset from the TFRecord file(s) in Cloud Storage.
   imageDataset = tf.data.TFRecordDataset(imageFilesList,
compression _type = 'GZIP')
   imageDataset = imageDataset.map(parse_image, num_parallel
_calls = 5)
   imageDataset = imageDataset.map(toTupleImage).batch(1)
# prediction
   print('Running predictions...')
   predictions = m.predict(imageDataset, steps = patches, verbose
= 1)# torch Tensor to TF tensor
   predictions[predictions >=0.5] = 1
   predictions[predictions<0.5] = 0
   print(predictions)
   print('Writing predictions...')
   print('Writing Tiff Image...')
   ab=[]
   ab2= []
   patches = 0
   x_buffer = int(kernel_buffer[0] / 2) # buffer is an extra piece of
the original size. Details can be found at https://developers.goo
gle.com/earth-engine/guides/tfrecord
   y_buffer = int(kernel_buffer[1] / 2)
   for predictionPatch in predictions:
      print('Writing patch ' + str(patches) + '...')
      predictionPatch = predictionPatch[x_buffer:x_buffer+KERNEL_
SIZE, y_buffer:y_buffer+KERNEL_SIZE]
      ab = predictionPatch[None,:, :,:]
      if patches == 0:
         ab2 = ab
      else :
         ab2 = tf.concat([ab2,ab], axis=0)
      patches += 1
   print('Writing Tiff Image...')
   #Export the prediction
   driver = gdal.GetDriverByName('GTiff')
   save_path = base_dir + out_image_base + '#tifname'
  rows=mixer['totalPatches']//mixer['patchesPerRow']*mixer
['patchDimensions'][1]
   cols=mixer['patchesPerRow']*mixer ['patchDimensions'][0]
   dataset = driver.Create(save_path,cols,rows,1, gdal.GDT_Byte)
   geotransform=mixer['projection']['affine']['doubleMatrix']
   index=[1,2,0,4,5,3]
   g=[]
   for i in index:
      g.append(geotransform[index[i]])
   dataset.SetGeoTransform(g)
   srs = osr.SpatialReference()
  srs.ImportFromEPSG(int(mixer['projection']['crs'].split(':')[-
1]))
   print(srs)
```

```
    dataset.SetProjection(srs.ExportToWkt())
  # predictions=np.load('./predict.npy')
    ab2 =ab2.numpy()
      p = ab2.reshape((-1, mixer['patchesPerRow'], KERNEL_SIZE,
  KERNEL_SIZE))
    raster = np.ones( (rows,cols ),dtype=np.uint8)
    for row in range(mixer['totalPatches']//mixer
  ['patchesPer Row']):
      for col in range(mixer['patchesPerRow']):
        r_index=row*KERNEL_SIZE
        c_index=col*KERNEL_SIZE

  raster[r_index:r_index+KERNEL_SIZE,c_index:c_index+KERNEL_
  SIZE]=p[row,col,:,:]
    dataset.GetRasterBand(1).WriteArray( raster )
  wc_image_base = 'UnetAWC'# wc_image_base in Data Preprocessing
  user_folder = 'data_path'
  doPrediction(wc_image_base , user_folder)
```

**[Python3.9] U-Net Classification Training**

```
import tensorflow as tf
import numpy as np
import os
from osgeo import gdal,osr
import matplotlib.pyplot as plt
# Sizes of the training and evaluation datasets.
TRAIN_SIZE =252
EVAL_SIZE =588
# Specify model training parameters.
BATCH_SIZE = 8
EPOCHS = 200
BUFFER_SIZE = 100
Bands = ['B2', 'B3', 'B4', 'B5', 'B6', 'B7',"ELEVATION","SLOPE",
"ASPECT"] #landsat+dem
#Bands = ['B2', 'B3', 'B4']#RGB
#Bands = ['B2', 'B3', 'B4', 'B5', 'B6', 'B7']#landsat
lable_name = ['label']
BANDs=Bands+lable_name #lable_name
# Specify the size and shape of patches expected by the model.
KERNEL_SIZE = 128 #
KERNEL_SHAPE = [KERNEL_SIZE, KERNEL_SIZE]
COLUMNS = [
 tf.io.FixedLenFeature(shape=KERNEL_SHAPE, dtype=tf.float32) for
k in Bands
]+[tf.io.FixedLenFeature(shape=KERNEL_SHAPE,dtype=tf.float32)]
# Create a TF object with the same size as the data
FEATURES_DICT = dict(zip(BANDs, COLUMNS)) #
# tf data analysis
def parse_tfrecord(example_proto):
  """The parsing function.
 Read a serialized example into the structure defined by FEATURES_
DICT.
```

```
  Args:
   example_proto: a serialized Example.
  Returns:
   A dictionary of tensors, keyed by feature name.
  """
 return tf.io.parse_single_example(example_proto, FEATURES_DICT)
def to_tuple(inputs):
   """Function to convert a dictionary of tensors to a tuple of
(inputs, outputs).
  Turn the tensors returned by parse_tfrecord into a stack in HWC
shape.
  Args:
   inputs: A dictionary of tensors, keyed by feature name.
  Returns:
   A tuple of (inputs, outputs).
  """
  inputsList = [inputs.get(key) for key in Bands]
  stacked = tf.stack(inputsList, axis=0)
  # Convert from CHW to HWC
  stacked = tf.transpose(stacked, [1, 2, 0])
  cls=inputs.get(lable_name[0])
  return stacked[:,:,:len(Bands)],cls
def get_dataset(pattern):
   """Function to read, parse and format to tuple a set of input
tfrecord files.
 Get all the files matching the pattern, parse and convert to tuple.
  Args:
   pattern: A file pattern to match in a Cloud Storage bucket.
  Returns:
   A tf.data.Dataset
  """
  glob = tf.io.gfile.glob(pattern)
  dataset = tf.data.TFRecordDataset(glob, compression_type =
'GZIP')
  dataset = dataset.map(parse_tfrecord, num_parallel_calls=5)
  dataset = dataset.map(to_tuple, num_parallel_calls=5)
  return dataset
def get_training_dataset(glob):
   """Get the preprocessed training dataset
  Returns:
   A tf.data.Dataset of training data.
  """
   # glob = 'gs://' + BUCKET + '/' + FOLDER + '/' + TRAINING_BASE +
'*'
   # glob = './drive/MyDrive/'+FOLDER+ '/' + TRAINING_BASE + '*'
   # glob = '/content/drive/Shareddrives/LCZ_drive/GBA_Test/*'
   dataset = get_dataset(glob)
   dataset = dataset.shuffle(BUFFER_SIZE).batch(BATCH_SIZE).
repeat()
   return dataset
glob1 = r'C:\Users\LAgO\Desktop\landslidedata\Unet\SIZE\128\128T\
step2\*'
training = get_training_dataset(glob1)
```

```
glob = r'C:\Users\LAgO\Desktop\landslidedata\Unet\SIZE\128\128V\
step2v\*'
val = get_training_dataset(glob)
demo=iter(training.take(1)).next()
print(demo)
#demov=iter(valing.take(1)).next()
#print(demov)
from tensorflow.python.keras import layers
from tensorflow.python.keras import models
def conv_block(input_tensor, num_filters):
  encoder = layers.Conv2D(num_filters, (3, 3), padding =
'same')(input_tensor)
  encoder = layers.BatchNormalization()(encoder)
  encoder = layers.Activation('relu')(encoder)
  encoder = layers.Conv2D(num_filters, (3, 3), padding =
'same')(encoder)
  encoder = layers.BatchNormalization()(encoder)
  encoder = layers.Activation('relu')(encoder)
  return encoder
def encoder_block(input_tensor, num_filters):
  encoder = conv_block(input_tensor, num_filters)
  encoder_pool = layers.MaxPooling2D((2, 2), strides =
(2, 2))(encoder)
  return encoder_pool, encoder
def decoder_block(input_tensor, concat_tensor, num_filters):
  decoder = layers.Conv2DTranspose(num_filters, (2, 2), strides =
(2, 2), padding='same')(input_tensor)
  decoder = layers.concatenate([concat_tensor, decoder], axis =
-1)
  decoder = layers.BatchNormalization()(decoder)
  decoder = layers.Activation('relu')(decoder)
  decoder = layers.Conv2D(num_filters, (3, 3), padding =
'same')(decoder)
  decoder = layers.BatchNormalization()(decoder)
  decoder = layers.Activation('relu')(decoder)
  decoder = layers.Conv2D(num_filters, (3, 3), padding =
'same')(decoder)
  decoder = layers.BatchNormalization()(decoder)
  decoder = layers.Activation('relu')(decoder)
  return decoder
def get_model():
  inputs = layers.Input(shape=[None, None, len(Bands)])
  encoder0_pool, encoder0 = encoder_block(inputs, 16)
  encoder1_pool, encoder1 = encoder_block(encoder0_pool, 32)
  encoder2_pool, encoder2 = encoder_block(encoder1_pool, 64)
  encoder3_pool, encoder3 = encoder_block(encoder2_pool, 128)
  encoder4_pool, encoder4 = encoder_block(encoder3_pool, 256)
  center = conv_block(encoder4_pool, 512)
  decoder4 = decoder_block(center, encoder4, 256)
  decoder3 = decoder_block(decoder4, encoder3, 128)
  decoder2 = decoder_block(decoder3, encoder2, 64)
  decoder1 = decoder_block(decoder2, encoder1, 32)
  decoder0 = decoder_block(decoder1, encoder0, 16)
  outputs = layers.Conv2D(1, (1, 1), activation =
```

```
'softmax')(decoder0)
    model = models.Model(inputs=[inputs], outputs=[outputs])
   model.compile(optimizer = 'adam', loss = 'binary_crossentropy',
metrics = ['Accuracy','Precision','Recall','AUC'])
    return model
m = get_model()
#MODEL_DIR = 'model_path'
mpath = r'save_model_path'
#m = tf.keras.models.load_model(MODEL_DIR)
cp_callback_save_best = tf.keras.callbacks.ModelCheckpoint(filepath
=mpath,monitor='val_Accuracy',verbose=1,save_best_only=True,
mode='max',period=10)
csv_logger = tf.keras.callbacks.CSVLogger('training.log',append
=False)
history =m.fit(
    x=training,
    epochs=EPOCHS,
    steps_per_epoch=int(TRAIN_SIZE / BATCH_SIZE),
    validation_data=val,
    validation_steps=EVAL_SIZE,
    callbacks=[cp_callback_save_best])
#Plotting
def print_history(history):
    plt.subplot(3, 1, 1)
    plt.plot(history.history['Accuracy'])
    plt.plot(history.history['val_Accuracy'])
    plt.plot(history.history['loss'])
    plt.plot(history.history['val_loss'])
    plt.legend(['Train_acc', 'Val_acc', 'Train_loss', 'Val_loss'])
    plt.subplot(3, 1, 2)
    plt.plot(history.history['recall'])
    plt.plot(history.history['val_recall'])
    plt.plot(history.history['precision'])
    plt.plot(history.history['val_precision'])
    plt.legend(['Train_recall','Val_recall','precision','vp'])
    plt.subplot(3, 1, 3)
    plt.plot(history.history['auc'])
    plt.plot(history.history['val_auc'])
    plt.legend(['Train_auc','Val_auc'])
    plt.title('Model accuracy&loss')
    plt.xlabel('Epoch')
    plt.show()
print(history.params)
print_history(history)
```

## [Python3.9] U-Net Classification Testing

```
import tensorflow as tf
import numpy as np
import os
from osgeo import gdal,osr
import matplotlib.pyplot as plt
BUFFER_SIZE = 100
```

```
Bands = ['B2', 'B3', 'B4', 'B5', 'B6', 'B7',"ELEVATION","SLOPE",
"ASPECT"] #landsat+dem
#Bands = ['B2', 'B3', 'B4']#RGB
#Bands = ['B2', 'B3', 'B4', 'B5', 'B6', 'B7']#landsat
lable_name = ['label']
BANDs=Bands+lable_name #lable_name
# Specify the size and shape of patches expected by the model.
KERNEL_SIZE = 128 #
KERNEL_SHAPE = [KERNEL_SIZE, KERNEL_SIZE]
COLUMNS = [
 tf.io.FixedLenFeature(shape=KERNEL_SHAPE, dtype=tf.float32) for
k in Bands
]+[tf.io.FixedLenFeature(shape=KERNEL_SHAPE,dtype=tf.float32)]
FEATURES_DICT = dict(zip(BANDs, COLUMNS))
# tf data analysis
def parse_tfrecord(example_proto):
  """The parsing function.
  Read a serialized example into the structure defined by FEATURES_
DICT.
  Args:
    example_proto: a serialized Example.
  Returns:
    A dictionary of tensors, keyed by feature name.
  """

 return tf.io.parse_single_example(example_proto, FEATURES_DICT)
def to_tuple(inputs):
   """Function to convert a dictionary of tensors to a tuple of
(inputs, outputs).
  Turn the tensors returned by parse_tfrecord into a stack in HWC
shape.
  Args:
    inputs: A dictionary of tensors, keyed by feature name.
  Returns:
    A tuple of (inputs, outputs).
  """
  inputsList = [inputs.get(key) for key in Bands]
  stacked = tf.stack(inputsList, axis=0)
  # Convert from CHW to HWC
  stacked = tf.transpose(stacked, [1, 2, 0])
  cls=inputs.get(lable_name[0])
  return stacked[:,:,:len(Bands)],cls
def get_dataset(pattern):
   """Function to read, parse and format to tuple a set of input
tfrecord files.
  Get all the files matching the pattern, parse and convert to tuple.
  Args:
    pattern: A file pattern to match in a Cloud Storage bucket.
  Returns:
    A tf.data.Dataset
  """
  glob = tf.io.gfile.glob(pattern)
  dataset = tf.data.TFRecordDataset(glob, compression_type =
'GZIP')
```

```
  dataset = dataset.map(parse_tfrecord, num_parallel_calls=5)
  dataset = dataset.map(to_tuple, num_parallel_calls=5)
  return dataset
def get_training_dataset(glob):
   """Get the preprocessed training dataset
  Returns:
   A tf.data.Dataset of training data.
  """
   dataset = get_dataset(glob)
   dataset = dataset.shuffle(BUFFER_SIZE).batch(BATCH_SIZE).
repeat ()
   return dataset
test_glob = r'testdata_path'
testdata = get_training_dataset(test_glob)
MODEL_DIR = r'model_path'
m = tf.keras.models.load_model(MODEL_DIR)
m.summary()
m.evaluate(testdata ,steps=100,verbose=1)
```

## References


Althuwaynee, O. F., Pradhan, B., & Lee, S. (2012). Application of an evidential belief function model in landslide susceptibility mapping. *Computers & Geosciences, 44*, 120–135.

Assilzadeh, H., Levy, J. K., & Wang, X. (2010). Landslide catastrophes and disaster risk reduction: A GIS framework for landslide prevention and management. *Remote Sensing, 2*(9), 2259–2273.

Behling, R., Roessner, S., Kaufmann, H., & Kleinschmit, B. (2014). Automated spatiotemporal landslide mapping over large areas using RapidEye time series data. *Remote Sensing, 6*(9), 8026–8055.

Belgiu, M., & Drăguţ, L. (2016). Random forest in remote sensing: A review of applications and future directions. *ISPRS Journal of Photogrammetry and Remote Sensing, 114*, 24–31.

Brardinoni, F., Slaymaker, O., & Hassan, M. A. (2003). Landslide inventory in a rugged forested watershed: A comparison between air-photo and field survey data. *Geomorphology, 54*, 179–196.

Breiman, L. (2001). Random forests. *Machine Learning, 45*, 5–32.

Chang, F. J., Chiang, Y. M., Tsai, M. J., Shieh, M. C., Hsu, K. L., & Sorooshian, S. (2014). Watershed rainfall forecasting using neuro-fuzzy networks with the assimilation of multi-sensor information. *Journal of Hydrology, 508*, 374–384.

Chen, T.-H.K., Prishchepov, A. V., Fensholt, R., & Sabel, C. E. (2019). Detecting and monitoring long-term landslides in urbanized areas with nighttime light data and multi-seasonal Landsat imagery across Taiwan from 1998 to 2017. *Remote Sensing of Environment, 225*, 317–327.

Chen, Z., Zhang, Y., Ouyang, C., Zhang, F., & Ma, J. (2018). Automated landslides detection for mountain cities using multi-temporal remote sensing imagery. *Sensors, 18*(3), 821.

Cheng, L., Li, J., Duan, P., & Wang, M. (2021). A small attentional YOLO model for landslide detection from satellite remote sensing images. *Landslides, 18*(8), 2751–2765.

Chou, C., Chiang, J. C., Lan, C. W., Chung, C. H., Liao, Y. C., & Lee, C. J. (2013). Increase in the range between wet and dry season precipitation. *Nature Geoscience, 6*(4), 263–267.

Chuang, Y. C., & Shiu, Y. S. (2018). Relationship between landslides and mountain development—Integrating geospatial statistics and a new long-term database. *Science of the Total Environment, 622*, 1265–1276.

Cui, P., Chen, X. Q., Zhu, Y. Y., Su, F. H., Wei, F. Q., Han, Y. S., Liu, H. J., & Zhuang, J. Q. (2011). The Wenchuan earthquake (May 12, 2008), Sichuan province, China, and resulting geohazards. *Natural Hazards, 56*, 19–36.

Cui, P., Guo, X., Jiang, T., Zhang, G., & Jin, W. (2019). Disaster effect induced by Asian water tower change and mitigation strategies. *Bulletin of Chinese Academy of Sciences, 34*(11), 1313–1321.
Cui, P., Wei, F. Q., He, S. M., You, Y., Chen, X. Q., Li, Z. L., Dang, C., & Yang, C. L. (2008). Mountain disasters induced by the earthquake of May 12 in Wenchuan and the disasters mitigation. *Journal of Mountain Science, 26*(3), 280–282.
Decuyper, M., Chávez, R. O., Lohbeck, M., Lastra, J. A., Tsendbazar, N., Hackländer, J., Herold, M., & Vågen, T. G. (2022). Continuous monitoring of forest change dynamics with satellite time series. *Remote Sensing of Environment, 269*, 112829.
Du, B., Ru, L., Wu, C., & Zhang, L. (2019). Unsupervised deep slow feature analysis for change detection in multi-temporal remote sensing images. *IEEE Transactions on Geoscience and Remote Sensing, 57*(12), 9976–9992.
Du, J., Glade, T., Woldai, T., Chai, B., & Zeng, B. (2020). Landslide susceptibility assessment based on an incomplete landslide inventory in the Jilong Valley, Tibet, Chinese Himalayas. *Engineering Geology, 270*, 105572.
Fan, X., Scaringi, G., Domènech, G., Yang, F., Guo, X., Dai, L., He, C., Xu, Q., & Huang, R. (2019). Two multi-temporal datasets that track the enhanced landsliding after the 2008 Wenchuan earthquake. *Earth System Science Data, 11*(1), 35–55.
Fang, B., Chen, G., Pan, L., Kou, R., & Wang, L. (2020). GAN-based Siamese framework for landslide inventory mapping using bi-temporal optical remote sensing images. *IEEE Geoscience and Remote Sensing Letters, 18*(3), 391–395.
Foga, S., Scaramuzza, P. L., Guo, S., Zhu, Z., Dilley, R. D., Jr., Beckmann, T., Schmidt, G. L., Dwyer, J. L., Hughes, M. J., & Laue, B. (2017). Cloud detection algorithm comparison and validation for operational Landsat data products. *Remote Sensing of Environment, 194*, 379–390.
Ghorbanzadeh, O., Blaschke, T., Gholamnia, K., Meena, S. R., Tiede, D., & Aryal, J. (2019). Evaluation of different machine learning methods and deep-learning convolutional neural networks for landslide detection. *Remote Sensing, 11*(2), 196.
Guo, X., Cui, P., Li, Y., Ma, L., Ge, Y., & Mahoney, W. B. (2016). Intensity–duration threshold of rainfall-triggered debris flows in the Wenchuan earthquake affected area, China. *Geomorphology, 253*, 208–216.
Guzzetti, F., Mondini, A. C., Cardinali, M., Fiorucci, F., Santangelo, M., & Chang, K. T. (2012). Landslide inventory maps: New tools for an old problem. *Earth-Science Reviews, 112*, 42–66.
Hölbling, D., Füreder, P., Antolini, F., Cigna, F., Casagli, N., & Lang, S. (2012). A semi-automated object-based approach for landslide detection validated by persistent scatterer interferometry measures and landslide inventories. *Remote Sensing, 4*(5), 1310–1336.
Hu, Z., Xu, H., Wang, C., Zhou, Z., Li, Z., & Li, C. (2018). A landslides detection method based on time series remote sensing images. *Spacecraft Recovery & Remote Sensing, 39*(2), 104–114.
Huang, R., & Li, W. (2014). Post-earthquake landsliding and long-term impacts in the Wenchuan earthquake area, China. *Engineering Geology, 182*, 111–120.
Jiang, W., Xi, J., Li, Z., Ding, M., Yang, L., & Xie, D. (2021). Landslide detection and segmentation using mask R-CNN with simulated hard samples. *Geomatics and Information Science of Wuhan University*. https://doi.org/10.13203/j.whugis20200692
Ju, Y., Xu, Q., Jin, S., Li, W., Dong, X., & Guo, Q. (2020). Automatic object detection of loess landslide based on deep learning. *Geomatics and Information Science of Wuhan University, 45*(11), 1747–1755.
Li, C., Wang, M., & Liu, K. (2018). A decadal evolution of landslides and debris flows after the Wenchuan earthquake. *Geomorphology, 323*, 1–12.
Li, C., Yi, B., Gao, P., Li, H., Sun, J., Chen, X., & Zhong, C. (2021). Valuable clues for DCNN-based landslide detection from a comparative assessment in the Wenchuan earthquake area. *Sensors, 21*(15), 5191.
Liu, C., & Chen, C. (2020). Research on the origins of geological disasters in China. *Geological Review, 66*(5), 1334–1348.

Liu, Y., Hu, C., Sun, C., Zhan, W., Sun, S., Xu, B., & Dong, Y. (2018). Assessment of offshore oil/gas platform status in the northern Gulf of Mexico using multi-source satellite time-series images. *Remote Sensing of Environment, 208*, 63–81.

Lv, Z. Y., Shi, W., Zhang, X., & Benediktsson, J. A. (2018). Landslide inventory mapping from bitemporal high-resolution remote sensing images using change detection and multiscale segmentation. *IEEE Journal of Selected Topics in Applied Earth Observations and Remote Sensing, 11*(5), 1520–1532.

Markham, B. L., Thome, K. J., Barsi, J. A., Kaita, E., Helder, D. L., Barker, J. L., & Scaramuzza, P. L. (2004). Landsat-7 ETM+ on-orbit reflective-band radiometric stability and absolute calibration. *IEEE Transactions on Geoscience and Remote Sensing, 42*(12), 2810–2820.

Mezaal, M. R., Pradhan, B., & Rizeei, H. M. (2018). Improving landslide detection from airborne laser scanning data using optimized Dempster–Shafer. *Remote Sensing, 10*(7), 1029.

Mwaniki, M. W., Agutu, N. O., Mbaka, J. G., Ngigi, T. G., & Waithaka, E. H. (2015). Landslide scar/soil erodibility mapping using Landsat TM/ETM+ bands 7 and 3 Normalised Difference Index: A case study of central region of Kenya. *Applied Geography, 64*, 108–120.

Mwaniki, M. W., Kuria, D. N., Boitt, M. K., & Ngigi, T. G. (2017). Image enhancements of Landsat 8 (OLI) and SAR data for preliminary landslide identification and mapping applied to the central region of Kenya. *Geomorphology, 282*, 162–175.

Parker, R. N., Densmore, A. L., Rosser, N. J., De Michele, M., Li, Y., Huang, R., Whadcoat, S., & Petley, D. N. (2011). Mass wasting triggered by the 2008 Wenchuan earthquake is greater than orogenic growth. *Nature Geoscience, 4*(7), 449–452.

Sameen, M. I., & Pradhan, B. (2019). Landslide detection using residual networks and the fusion of spectral and topographic information. *IEEE Access, 7*, 114363–114373.

Stumpf, A., & Kerle, N. (2011). Combining Random Forests and object-oriented analysis for landslide mapping from very high resolution imagery. *Procedia Environmental Sciences, 3*, 123–129.

Wang, W., Godard, V., Liu-Zeng, J., Scherler, D., Xu, C., Zhang, J., Xie, K., Bellier, O., Ansberque, C., De Sigoyer, J., & ASTER Team. (2017). Perturbation of fluvial sediment fluxes following the 2008 Wenchuan earthquake. *Earth Surface Processes and Landforms, 42*(15), 2611–2622.

Wang, Y., Fang, Z., Wang, M., Peng, L., & Hong, H. (2020). Comparative study of landslide susceptibility mapping with different recurrent neural networks. *Computers & Geosciences, 138*, 104445.

Wei, L. W., Huang, C. M., Chen, H., Lee, C. T., Chi, C. C., & Chiu, C. L. (2018). Adopting the $I_3$–$R_{24}$ rainfall index and landslide susceptibility for the establishment of an early warning model for rainfall-induced shallow landslides. *Natural Hazards and Earth System Sciences, 18*(6), 1717–1733.

Xing, L., & Niu, Z. (2019). Mapping and analyzing China's wetlands using MODIS time series data. *Wetlands Ecology and Management, 27*, 693–710.

Xu, C., Xu, X., Shen, L., Yao, Q., Tan, X., Kang, W., Ma, S., Wu, X., Cai, J., Gao, M., & Li, K. (2016). Optimized volume models of earthquake-triggered landslides. *Scientific Reports, 6*(1), 29797.

Xu, C., Xu, X., Yao, X., & Dai, F. (2014). Three (nearly) complete inventories of landslides triggered by the May 12, 2008 Wenchuan Mw 7.9 earthquake of China and their spatial distribution statistical analysis. *Landslides, 11*, 441–461.

Xu, Q., Dong, X., & Li, W. (2019). Integrated space-air-ground early detection, monitoring and warning system for potential catastrophic geohazards. *Geomatics and Information Science of Wuhan University, 44*(7), 957–966.

Yang, M., Li, S., & Feng, Z. (2018). Post-seismic landslide extraction by combining texture analysis and emissivity estimating. *Journal of Remote Sensing, 22*, 212–223.

Yi, Y., & Zhang, W. (2020). A new deep-learning-based approach for earthquake-triggered landslide detection from single-temporal RapidEye satellite imagery. *IEEE Journal of Selected Topics in Applied Earth Observations and Remote Sensing, 13*, 6166–6176.

Zhang, P., Deng, Q., Zhang, Z., & Li, H. (2013). Active faults, earthquake hazards and associated geodynamic processes in continental China (in Chinese). *Scientia Sinica Terrae, 43*, 1607–1620.

Zhao, R., Wang, G., Yu, X., Sun, X., Bao, Y., Xiao, G., Gan, W., & Shen, S. (2020). Rapid land subsidence in Tianjin, China derived from continuous GPS observations (2010–2019). *Proceedings of the International Association of Hydrological Sciences, 382*, 241–247.

Zhao, Y., Wang, R., Jiang, Y., Liu, H., & Wei, Z. (2019). GIS-based logistic regression for rainfall-induced landslide susceptibility mapping under different grid sizes in Yueqing, Southeastern China. *Engineering Geology, 259*, 105147.

Zhong, C., Li, C., Gao, P., & Li, H. (2021). Discovering vegetation recovery and landslide activities in the Wenchuan earthquake area with Landsat imagery. *Sensors, 21*(15), 5243.

# Chapter 5
# Deep Learning Based Remote Sensing Monitoring of Landslide

**Abstract** Landslide is a common natural disaster that can easily cause significant economic loss and threaten human life. It is of great importance to identify landslide accurately for disaster prevention and mitigation. Deep learning (DL) has been proved to be effective in object detection tasks and has been applied in a variety of study realms. This chapter proposed a DL-based framework to identify landslides. Faster R-CNN is used in this research, with a graph convolutional neural network (GCN) module to improve the model performance. A landslide dataset constructed based on Gaofen-2 satellite remote sensing images is used for this study. The results indicate the framework can successfully identify most landslides in the dataset. And Faster R-CNN equipped with GCN shows a better performance than the original framework.

**Keywords** Landslide · Remote sensing · Deep learning · Object detection

## 5.1 Introduction

### *5.1.1 Background and Significance*

Landslide is a natural disaster that is formed when the soil or rock mass on the slope are affected by natural factors (erosion, immersion, earthquake, etc.), which slide down along the sliding surface of the slope under the effect of gravity. China has a vast territory with diverse terrains, so it suffers from widely distributed and frequent geological disasters. According to the Ministry of Natural Resources of the People's Republic of China, 4772 geological disasters were reported in China in 2021, leading to 80 deaths and 11 missing cases, and causing the direct economic loss of 3.2 billion RMB in total. The main disaster types were landslide with 2335 cases, collapse with 1746 cases, and debris flow with 374 cases. Landslide took the dominant space with 48.9% of total geological disaster. On the other hand, it was also mentioned 905 geologic disasters were successfully forecasted, avoiding 25,528 possible casualties and economic loss of 1.35 billion RMB. Therefore, developing early warning systems for the geological disasters can effectively lower their risks.

 
W. Chen et al., *Intelligent Interpretation for Geological Disasters*,
https://doi.org/10.1007/978-981-99-5822-1_5

In the landslide-related researches, landslide identification is often the basis of other works. The identification of historical landslides can indicate the distribution of landslides so as to assess the landslide susceptibility. And the identification of early landslides can prevent undamaged, but unstable slopes. Satellite remote sensing data is widely used in the studies of various geological disasters in recent decades. In the early stages, landslides were interpreted manually from remote sensing images according to the experience and professional knowledge of the staff. This method was inefficient and failed to identify disasters in real time. With the development of computer technology, semi-automatic identification method of human–computer interaction (HCI) was then used in this realm, by manually setting criteria for identifying landslides, and then identifying landslides with computer algorithms. Zhang and Zhao (2017) developed the landslide recognition criteria based on the texture, shape and spectrum for landslide detection in combination with the object-oriented technology. Huang et al. (2018) optimized the extraction method of parameters and used multi-scale segmentation algorithm to identify landslides. Chen et al. (2011) combined the CCD and SAR image data to develop a HCI interpretation algorithm for detecting landslide area. Huang (2009) designed landslide detection criteria with context relations and features in remote sensing images. However, this human-involved method still cannot satisfy the demand of rapid landslide detection due to the increasing amount of landslide data.

Deep learning (DL) is a data-driven method that has been utilized in many fields in recent years (Szegedy et al., 2013). It can learn the potential features without prior knowledge of the specific task. In order to timely process and make full use of landslide data, deep learning has become a research hotspot of landslide detection methods.

### *5.1.2 Research Overview*

#### Status of Landslide Detection

With the improvement of computer computing power and the rapid development of deep learning network, these automatic data-driven methods are more usually used to process massive data. However, due to the formation and final form of landslide, as well as the complex and changeable surrounding environment, there are still many challenges in landslide detection by deep learning model. There are four main landslide identification methods:

(1) Traditional artificial visual interpretation method. Early professional interpreters identified landslide based on the geometric features of remote sensing images, such as color, tone, texture and shadow, and then analyzed and inferred the landslide area based on experience and relevant geological knowledge combined with other data. This method is reliable, but very inefficient.
(2) Change detection method. Change detection was developed gradually in the 1980s to identify the processing methods of the same surface information

obtained from remote sensing images and geographical data at different times. This method is suitable for remote sensing data with time series characteristics, and the landslide can be identified by the changed area. Behling et al. (2014) used multi-temporal Rapid Eye satellite remote sensing images to detect the slope according to the different growth rates of vegetation in slope region and non-slope region to the initial state in multiple time series images.

(3) Machine learning method. It is a method to use feature classifier to identify the landslide according to the extracted various related features from remote sensing data. There are some widely used algorithms for classifier development, such as Bayesian, logistic regression, support vector machine (SVM) and artificial neural network (ANN). Although this method is more automatic than the first two methods, it needs a lot of time to extract features and debug hyper parameters to get a more reasonable algorithm effect. Li et al. (2019) used BP neural network model fusing with other Bootstrap algorithms to predict landslide displacement interval. Guo et al. (2016) constructed the automatic interpretation landslide model by transfer learning feature and SVM. Chen et al. (2017) studied the effects of different data on landslide sensitivity maps using Logistic Regression and ANN models. Pham et al. (2017) combined the machine learning integration method and multilayer perceptron neural network to evaluate the sensitivity of landslide.

(4) Deep learning method. Compared with the aforementioned methods, the DL method does not require manual construction and feature selection in landslide data processing (Chen et al., 2021). In addition, deep learning can handle larger amount of data and adapt to a large range of landslide identification. Convolutional neural network (CNN) is mainly used in landslide detection because of its parallel learning of multi features and translational invariance. DL-based target detection algorithms are mainly divided into two categories. One is two-stage detection algorithm based on regional suggestion box, such as R-CNN, SPP-NET, Fast R-CNN and Faster R-CNN, etc. The other is regression-based single-stage detection algorithms, such as SSD, YOLO, RetinaNet, etc. Ju et al. (2020) used Mask R-CNN to explore the feasibility of loess landslide target detection. Shi et al. (2022) used YOLACT (You Only Look at Coefficients), a single stage instance segmentation network, for automatic multi-task identification of historical loess landslides.

### Status of Target Detection Methods

The traditional target detection methods rely on the image features, which are generally selected and designed artificially, and use sliding window machine and SVM classification algorithm to achieve detection. To be specific, the candidate region is selected by sliding window or region selection algorithm to determine the location of the target. Then the features of selected candidate regions are extracted artificially. Finally, the extracted features are input into the classifier to obtain the detection result. In the process of artificial feature extraction, due to the different detection targets, complex background, image shooting angle and distance, etc., the extraction result can be easily affected, so it is a time-consuming and important work. As deep learning has made great progress in the field of target detection, nowadays target detection algorithms that rely on image features are gradually being replaced by DL-based methods. DL-based target detection algorithms can be divided into two categories according to the network structure characteristics. The first one is two-stage target detection algorithm. The algorithm is implemented by two stages: (1) firstly from the target image select a number of candidate regions, and then use CNN for the feature extraction of candidate areas; (2) Secondly the interested candidate areas will be pooling mapped to feature maps to locate and decide that category. The other one is the single-stage algorithm that can directly output the classification and location results of targets.

In 2014, Girshick et al. (2014) proposed the two-stage target detection algorithm called R-CNN. R-CNN generates 1–2 k candidate regions through sliding window search, then extracts image features for each generated candidate region. These feature vectors were input into SVM classifier to determine categories, and regressor was used to correct the positions of candidate boxes. It is the framework prototype of target detection algorithm based on CNN. Comparing with the earlier target detection algorithm, the accuracy of R-CNN is much higher. However, due to the large number of overlapping boxes generated in the sliding window search, the computation of feature extraction network is very large, which leads to slow training and detection speed. GPU acceleration cannot be used due to algorithm design defects. Therefore, He et al. (2015) proposed SPPNet. SPPNet has a spatial pyramid pooling structure, which can accommodate any input image size. In this case, the whole image can be directly input into the network to extract the features of candidate regions at one time.

Fast R-CNN proposed in 2015 (Girshick, 2015) is also an optimization of R-CNN. The method of selective search is adopted to pre-process and extract boundary boxes to improve the speed of R-CNN. The feature extraction network was designed to an ROI (Region of Interest) pooling layer, and only one scale is required to divide candidate regions into a fixed number of zones. In addition, Fast R-CNN abandons SVM classifier and merges regressor and classifier by using full connection layer, so that they can be trained simultaneously in the network. The total loss is composed of classification loss and regression coordinate loss by weighted operation. This greatly reduces the computing amount of Fast R-CNN in the feature extraction network, eliminates the need to use other cache space to store features, reduces the complexity of training, and improves the detection accuracy.

Fast R-CNN needs firstly to conduct selective search to find all pre-selected boxes, this exhaustive search is still time-consuming. Ren et al. (2015) proposed the landmark Faster R-CNN model. The method is to design a neural network that can extract edges. With the help of convolutional network, the required pre-selection boxes can be automatically generated, which greatly reduces the search time of candidate boxes. It is the first end-to-end and nearly real-time detection algorithm.

Even have been improved a lot, the abovementioned two-stage detection algorithms still limit the speed of the algorithm to a certain extent. Therefore, researchers also try to design and improve the single-stage target detection algorithm, by which the position and category of objects in the image can be directly obtained from the output end when the images are input to the target detection network. In 2016, Redmon et al. (2016) proposed the YOLO network model. In this model, images are scaled to a uniform size and divided into fixed numbers of grids. Each grid unit is considered as the potential location of the target, and the 4-dimensional coordinate information of the rectangular box and the probability of the category are predicted. YOLO has a fast running speed, but due to its rough mesh division of images, it usually leads to significant object positioning errors. Liu et al. (2016) proposed the SSD network to carry out target detection purposefully on feature maps of various scales. Small-size targets were detected by large-scale feature maps containing more detailed features, and large-size targets were detected by small-scale feature maps containing more global semantic information. Meanwhile, SSD also uses the anchor box algorithm of Faster R-CNN to generate anchor boxes for different scales. These improving measures solve the problems of YOLO to some extent. However, for the single-stage detector, it is easy to produce a large number of negative samples, resulting in class imbalance during training. Lin et al., (2017a, 2017b) then proposed RetinaNet to improve SSD. Focalloss is introduced to allow the model to focus on the loss of the target itself in the training process.

Overall, the single-stage detection algorithms have higher detection speed, but the detection accuracy is lower than that of the two-stage detection algorithms. In the algorithm, the accuracy rate of Faster R-CNN is higher and the missed detection rate is lower (2017). All the above DL-based target detection algorithms extract image features with CNN. In fact, there may be potential relationship between features. Graph Convolutional Networks (GCN) can aggregate the information of surrounding neurals and extract more features. It is widely used in natural language processing and processing some non-Euclidean structure data. Introducing GCN in landslide target detection could be a potential way to improve the landslide identification efficiency.

**Existing Problems of Landslide Detection**

(1) Insufficient degree of automation. So far, the traditional geological survey is still the main method for landslide identification, but manual survey alone cannot identify the hidden landslides and landslides in unreachable regions.
(2) There are very few available landslide datasets and their data storing forms are not standardized, which make it hard to indicate the benefit of DL-based methods.
(3) As complex and diverse ground surface objects can show similar characteristics in the spectrum of satellite remote sensing, it is difficult to extract landslide features correctly, which makes the landslide target detection algorithm has low universality.
(4) Due to environmental factors, the distinction between landslides and non-landslides is reduced in the high coverage area. For example, the bedrock outcrop makes the features of landslide similar to those of other areas, such as surface roughness. Moreover, vegetation cover and mountain shadow can affect the landslide morphological feature, which can also be influenced by human activities.

### 5.1.3 Research Object and Contents

The purpose of this chapter is to construct a target detection model by integrating the existing target detection algorithm and graph convolution algorithm, and then use the model to identify landslides in remote sensing images and evaluate the model performance.

The content includes the processing of landslide dataset, the introduction of the landslide detection algorithm used in this research, the location of the graph convolution fusing into the target detection algorithm and a variety of different map construction methods, the parameter adjustment of the model, and finally the test and evaluation of the model. The flow chart is shown in Fig. 5.1.

(1) Firstly, remote sensing images of landslide area are obtained from multi-source high-resolution remote sensing satellites, which have been confirmed in ground survey. Based on the ground survey results, the landslide locations are determined by visual interpretation. Finally, the landslide images are pre-processed and standardized.
(2) Faster R-CNN and GCN are fused in many forms to explore a most suitable mapping method for the landslide data and construct the final model.
(3) Train and test the proposed model and evaluate the model performance.

The arrangement of each section is: In Sect. 5.2, Faster R-CNN and GCN are introduced. Section 5.3 describes the construction process of the landslide detection model based on Faster R-CNN and GCN. Section 5.4 describes the results of proposed landslide detection model and makes comparison with other models. Section 5.4 summarizes the research and illustrates the perspectives for landslide detection in future researches.

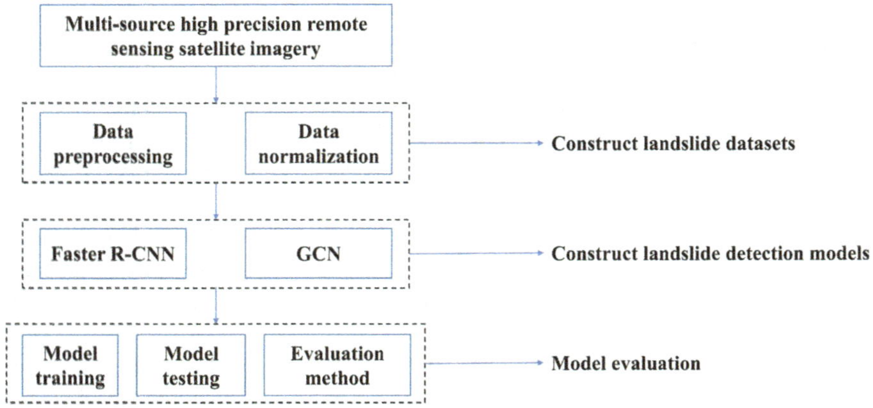


**Fig. 5.1** Flow chart of the DL-based landslide detection method

## 5.2 Related Principles and Techniques

### *5.2.1 Faster R-CNN Model*

Faster R-CNN is the third model of the R-CNN series and improved from R-CNN and Fast R-CNN, which also won in many projects of the COCO Competition in 2015.Compared with R-CNN, Fast R-CNN abandons the original method of separating SVM classification and border coordinate regression, and completes the classification and regression work after feature extraction in a unified network to accelerate the training speed. The Faster R-CNN can be regarded as the Fast R-CNN which improves the selective search algorithm with Region Proposal Network (RPN) to obtain candidate regions. Combination of candidate region generation and feature extraction allows the network to share the features, thereby improve detection speed. The network architecture of Faster R-CNN is shown in Fig. 5.2.

The Faster R-CNN network includes four parts:

(1) Convolutional layers. The input landslide images are pooled by convolution kernel to obtain image features. After the feature map is obtained, it will be input into the RPN network and the fully connected network.
(2) Region Proposal Network. After obtaining the feature map, the probability of foreground and background of each anchor point and the coordinate of the boxes are calculated to obtain the candidate region.

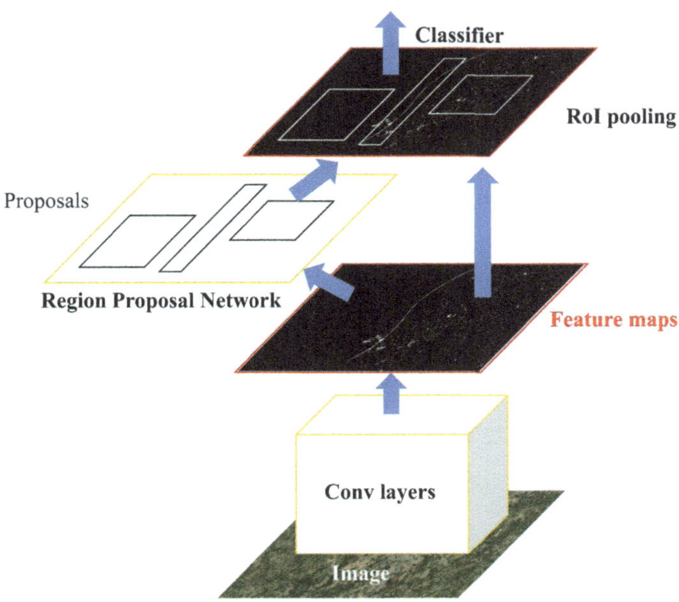


**Fig. 5.2** Faster R-CNN network structure, modified from Ren et al. (2015)

(3) Region of Interests (RoI) pooling layer. The location information of the region proposal is combined with the feature map to normalize the size of candidate regions for subsequent input of the full connection layer.
(4) Classifier layer. The feature map normalized by ROI pooling layer is used for the regression of category and accurate coordinate of detection box.

The most significant contribution of Faster R-CNN lies in the anchor box mechanism when RPN network is used to acquire candidate boxes. Anchor means the candidate box will have fixed sizes. This mechanism generates 9 anchors with different sizes for each point in the feature map. These anchors are generated by permutations and combinations of three scales (128, 256, 512) and three aspect ratios (1:1, 1:2, 2:1), as shown in Fig. 5.3, to cover objects of different sizes.

Similar to the sliding window, these 9 anchors will translate on the feature map. For an image with height H and width W, the number of generated anchors can be calculated by Eq. (5.1):

$$Sum_{anchor} = \left\lfloor \frac{H}{stridelength} \right\rfloor \times \left\lfloor \frac{W}{stridelength} \right\rfloor \times 9 \tag{5.1}$$

where stride length is the multiple of samplings of feature extraction network.

The input of RPN network includes not only feature map, but also the ground truth box (GT box) derived from the dataset. The overlapping area between labeled GT box and a series of anchors obtained according to Eq. (5.1) is calculated. Then each anchor can obtain an Intersection over Union (IoU), which is expressed by Eq. (5.2):

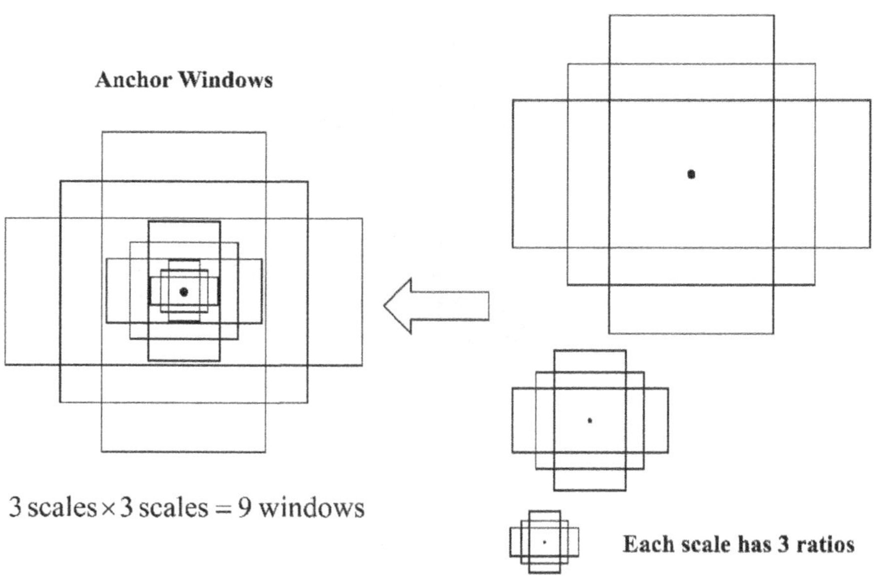


**Fig. 5.3** Anchor generation diagram

$$IoU = \frac{A \cap B}{A \cup B} \tag{5.2}$$

where A represents the prediction box generated by anchor, and B represents GT box. The confidence of each anchor can be obtained by calculating the area ratio of the intersection of prediction box and GT box to the area value of their union. The result can be used to determine whether the anchor contains a target.

When the Faster R-CNN needs to classify the foreground and background of anchor, it assigns a binary label to each anchor, which is used to identify whether the anchor needs to recognize the target category. When the anchor contains the target, it is the foreground. The binary label is defined as:

$$P_i^* = \begin{cases} 0 \ IoU > 0.7 \\ 1 \ IoU < 0.3 \end{cases} \tag{5.3}$$

In one image, when there is a maximum IoU value > 0.7 between the anchor and any GT box, the anchor is considered as a positive anchor, which means the foreground contains target. When IoU between the anchor and any GT box is lower than 0.3, it is considered as negative anchor, and the background contain no target. Anchors with IoU in 0.3 and 0.7 will be dropped. Using Softmax in RPN for the binary classification of foreground and background, the probability of each category can be calculated by using the following equation:

$$P_j = \frac{e_j^o}{\sum_k e^{o_k}}, j = 1, \ldots, k \tag{5.4}$$

where, $P_j$ is the category probability, $k$ is the number of categories, $j$ represents a specific category.

RPN also needs to perform regression of bounding box position when classifying foreground and background. The regression algorithm of bounding box position is shown in Fig. 5.4.

Box is generally represented by four-dimensional vectors $(x, y, w, h)$, which includes the center point coordinate of anchor and the width and height of box. When the four-dimensional vector of the original anchor candidate box regresses to the GT box, it means to look for a mapping relationship that allows the following equation to be established:

$$f(P) = \hat{G} \approx G \tag{5.5}$$

where, P represents the four-dimensional vector $(P_x, P_y, P_w, P_h)$ of the original anchor candidate box. $\hat{G}$ represents the four-dimensional vector $(\hat{G}_x, \hat{G}_y, \hat{G}_w, \hat{G}_h)$ of the prediction box and $G$ represents the four-dimensional vector $(G_x, G_y, G_w, G_h)$ of GT box.

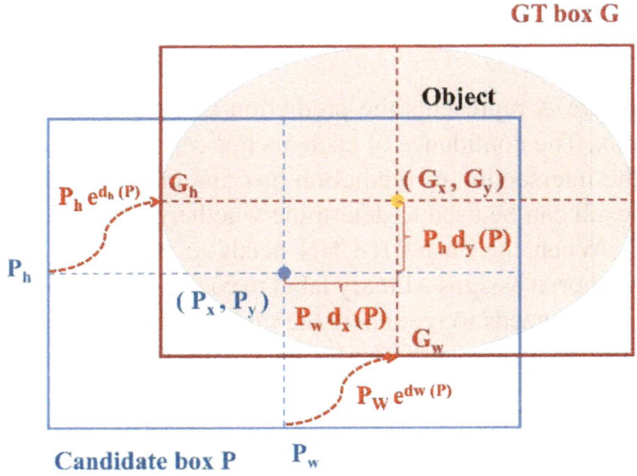


**Fig. 5.4** Bounding box position regression algorithm

This mapping relationship $f$ is actually the translation and scaling transformation of the original candidate box. During the transformation process, it keeps approaching the position and size of the real labelled GT box. Translation can be expressed as:

$$\Delta x = P_w d_x(P) \tag{5.6}$$

$$\Delta y = P_h d_y(P) \tag{5.7}$$

The translation $(\Delta x, \Delta y)$ can be obtained by Eq. (5.6) and Eq. (5.7). Based on the translation, the central point of the anchor of the original candidate box can translate following the axis to obtain the central point coordinate $\left(\hat{G}_x, \hat{G}_y\right)$ of the prediction box, which can be expressed as:

$$\hat{G}_x = \Delta x + P_x \tag{5.8}$$

$$\hat{G}_y = \Delta y + P_y \tag{5.9}$$

After the center position of the prediction box is determined, the size of the prediction box needs to be scaled. The scale $(S_w, S_h)$ is calculated by the following equations:

$$S_w = \exp(d_w(P)) \tag{5.10}$$

$$S_h = \exp(d_h(P)) \tag{5.11}$$

After the scaling ratio is obtained from Eq. (5.10) and Eq. (5.11), the translated anchor candidate box is scaled, which is closer to GT box in window scale, and the width and height of the prediction box can be finally obtained from:

$$\hat{G}_w = P_w S_w \tag{5.12}$$

$$\hat{G}_h = P_h S_h \tag{5.13}$$

The calculation of offset $(d_x, d_y)$ and scale $(d_w, d_h)$ factor in the abovementioned equations for the calculation of position and size of prediction box can be achieved with:

$$d_x = \frac{G_x - P_x}{P_w}, \quad d_y = \frac{G_y - P_y}{P_y} \tag{5.14}$$

$$d_w = \log\left(\frac{G_w}{P_w}\right), \quad d_h = \log\left(\frac{G_h}{P_h}\right) \tag{5.15}$$

where $d_x$, $d_y$ calculations divided by the width and height is to calculate in the proportion of translation to cater for images of different sizes. As scale factor must be greater than zero and *exp* is used to avoid the scale factor being negative, so log is used to calculate the $d_w$ and $d_h$.

The principle of RPN is shown in Fig. 5.5. The feature map is classified and regressed respectively through two $1 \times 1$ convolution operations. In the classification part, the number of channels of the feature map is 18, which is composed of 9 anchors with different sizes and 2 categories. It can determine whether the anchor is a foreground. In the regression part, the number of channels change to 36 after $1 \times 1$ convolution, which is formed by 9 anchors and 4D position vectors of each anchor. Information of all candidate regions is input into the RoI pooling layer to obtain the candidate regions in the feature map. Since the size of prediction boxes of candidate regions is different but the input size of full connection layer is fixed, the RoI pooling is used to normalize these candidate regions into a fixed size. In the full connection layer, softmax function is used again to determine the target category, and position of the prediction box is regressed to obtain a more accurate position.

### 5.2.2 *Graph Convolutional Network*

Bruna et al. (2013) generalized CNN to image data and proposed GCN. GCN can be divided into spectral domain and spatial domain based on its update mode. Spatial convolution is similar to the direct convolution on image pixels and is similar to the application of convolution in DL. The convolution based on spectral domain can also be called frequency domain convolution, which uses the Graph Fourier

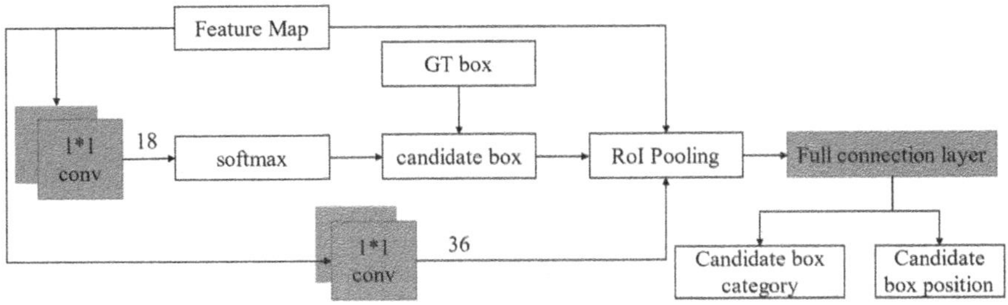


**Fig. 5.5** Principle of RPN

Transform in the graph theory for the convolution operation on image data. With the eigenvalue and eigenvector of the Laplace matrix, the Laplace operator of the graph in the frequency domain can be calculated. The equation of the graph convolution is obtained by analogy with the convolution in the Euclidean space in the frequency domain. Finally, the convolution operation on the graph is defined.

Generally, a graph is expressed as $G = \{V, E\}$, where $V$ is a set of $K$ graph nodes and $E$ is a set of graph edges. For a graph with $N$ nodes, the feature dimension of each node is $C$, then all node features will form the feature matrix $X$, whose dimension is $N \times C$. The relationship between nodes is represented by the $N \times N$ adjacency matrix $A$. The basic graph convolution equation can be expressed as:

$$x_i^{l+1} = \sigma\left(\sum_{j \in N_i} \frac{1}{c_{ij}} x_j^l W^l\right) \tag{5.16}$$

where, $\overrightarrow{x_i}^l$ and $\overrightarrow{x_i}^{l+1}$ are the representations of node $i$ before and after $l$ convolution layer, $N_i$ is the adjacent node of $i$, $W^l$ is the weight of the $l$ convolution layer, and $\sigma$ is the activation function of nonlinear transformation. $c_{ij}$ represents the normalized parameter, which can be obtained by the calculation of normalized adjacency matrix $D^{-\frac{1}{2}}AD^{-\frac{1}{2}}$. $D$ represents the degree matrix. Equation (5.16) can be transformed to Eq. (5.17) to represent the propagation between layers:

$$X^{l+1} = \sigma\left(\hat{A}X^l W^l\right) \tag{5.17}$$

where, $\hat{A} = \tilde{D}^{-\frac{1}{2}}\tilde{A}\tilde{D}^{-\frac{1}{2}}$, $\tilde{A} = \mathrm{A} + \mathrm{I}$, $I$ is the identity matrix, and $\tilde{D}$ is the degree matrix of $\tilde{A}$, which is for the normalization and symmetry of $\tilde{A}$. The diagonal of the adjacency matrix is 0, so the feature of its own node will be ignored in convolution. The function of the identity matrix is to make the diagonal be 1, so that its own feature will not be ignored.

## 5.3 Model Construction

### 5.3.1 Graph Convolutional Layer

A feature map can be defined as $X \in \mathbb{R}^{N \times C}$, where C is the number of feature channels and N is the number of pixels on each channel, which equals to the height H multiplies the width W. In conventional CNN, information is only exchanged in a small neighbor area. Acquisition of information depends on the size of the convolution kernel (as shown in Fig. 5.6, a 3 × 3 convolution kernel). To capture information between features with long distance, many convolution layers are needed. It worth mentioning that although the pooling layer can expand the receptive field, it still leads to the loss of information as this will reduce the number of features.

In Fig. 5.6, CNN convolution kernel learns a weight $\vec{a}$ for each feature point in part of the region contained by the kernel, and then the weights are integrated by multiplying the transformation matrix $W$. Differently, GCN extends the convolution operation of CNN to the data with graphic structure by constructing an adjacency matrix, as shown in Fig. 5.7.

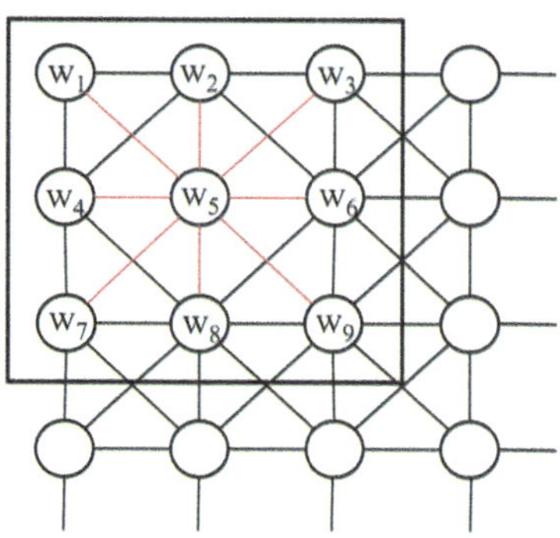


**Fig. 5.6** CNN convolution kernel, modified from Szegedy et al. (2013)

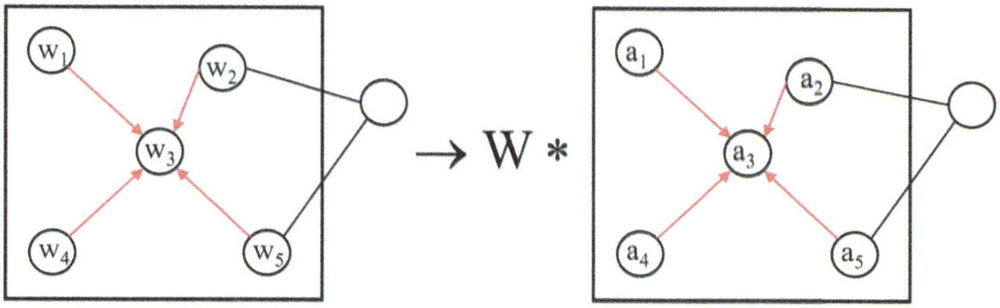


**Fig. 5.7** Graph convolution kernel, modified from Bruna et al. (2013)

Compared with the limited receptive field of CNN, GCN can expand the receptive field and capture context information. GCN was originally used in knowledge graph, but in recent years, it has been widely used in image feature generation. As an effective and efficient module and unlike the traditional convolution, graph convolution not only learns the neighbor neurons, but also allows remote information exchange in an individual layer. Since the parameter matrix $W$ of each neuron is the same, the weights are shared. Each aggregation is locally joined, which allows GCN to have less parameters. In addition, the receptive field of GCN is proportional to the number of convolutional layers. For neurons in the first layer, it contains information of their adjacent first-order neurons, while neurons in the second layer will aggregate information of second-order adjacent neurons, enabling information exchange with neurons with long distances.

The premise to use graph convolution is to design a connection mode to form topological graph, so as to obtain adjacency matrix, and then use Eq. (5.17) for convolution. Therefore, designing the structure of the graph is crucial for information transmission. Therefore, we define a network that can automatically learn the adjacency matrix according to the input feature map $X$. The generation process of the adjacency matrix is shown in Fig. 5.8.

According to the input tensor $X$, it is convolved twice respectively. The result of one convolution is stretched in the x direction, and that of the other convolution is stretched in the Y direction. The final adjacency matrix is formed by the multiplication of the two results, which can be expressed by the equation:

$$\tilde{A}_s = H_x(X)H_y(X) + I_C \tag{5.18}$$

where $H_x$ represents the result of convolution stretched in the x direction, $H_y$ represents the result of convolution stretched in the y direction, $\tilde{A}_s$ is spatial adjacency matrix, $I_c$ is the identity matrix with the size of $C \times C$.

Then according to formula (6.17), graph convolution operation can be defined as:

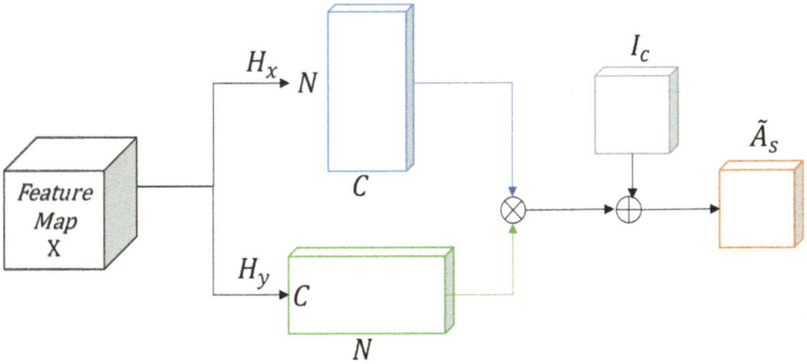


**Fig. 5.8** Adjacency matrix generation process

$$X^{(l+1)} = \sigma\left(\tilde{D}_s^{-\frac{1}{2}} \tilde{A}_s \tilde{D}_s^{-\frac{1}{2}} \left(X^T\right)^{(l)} W_s\right) \tag{5.19}$$

The graph convolution module is using the three-layer graph convolution to compute the representation in the feature space. Relu activation function is used in the first two layers. At the end of three-layer graph convolution, it is reshaped to the size of the original feature map and delivered to the subsequent network for following process.

### 5.3.2 Feature Pyramid Network

In the Faster R-CNN target detection, using the last high-level feature map of feature extraction network for the box regression and prediction may have a good effect for large objects, but small objects may be overlooked. Because, for small objects, the model often needs multiple downsamples in order to obtain high-level features. However, when the convolutional pooling is carried out in the last layer, the semantic information of small objects may be lost in the pooling process due to the small objects are presented in fewer pixels. Image pyramid can be used to deal with multi-scale problems, but the computation amount of the algorithm is considerable. In this case, Feature Pyramid Network (FPN) is introduced in this study (Lin, et al., 2017a, 2017b).

The main contribution of feature pyramid is to fuse the low-level feature information with the high-level semantic information, so both high-level feature and small object information can be obtained. The fusion methods of FPN for different levels can be summarized as two processes: bottom-up extraction of feature kernel and top-down fusion of features. In the bottom-up process, common CNN network is used as the backbone network to extract features and generate feature maps in different sizes. ResNet50 (He et al., 2016) is used as the backbone in this research. When the image is input into ResNet50, the last block of each residual module can output a feature map in a specific size, and the module with the same output size is called a stage. Five feature maps can be output in total. In this experiment, feature maps of the output layer from conv2 to conv5 are used, denoted as $\{C_2, C_3, C_4, C_5\}$. Their resolutions were halved in order, with 256, 512, 1024 and 2048 channels, respectively. The feature layer after fusion is denoted as $\{P_2, P_3, P_4, P_5, P_6\}$, and the fusion process is expressed by the following equation:

$$P_i = f\left(C_i, C_{i+1,}, \ldots, C_{i+n}\right) \tag{5.20}$$

where, $i$ is the output layer of ResNet, $n$ is the number of feature maps with different sizes, and $f$ is the fusion mode of feature layer.

In the top-down process, due to the different number of channels in each stage, $1 \times 1$ convolution is needed to unify the dimensions of stage to 256. In addition, as the resolution of each layer is halved successively, the feature resolution of the upper

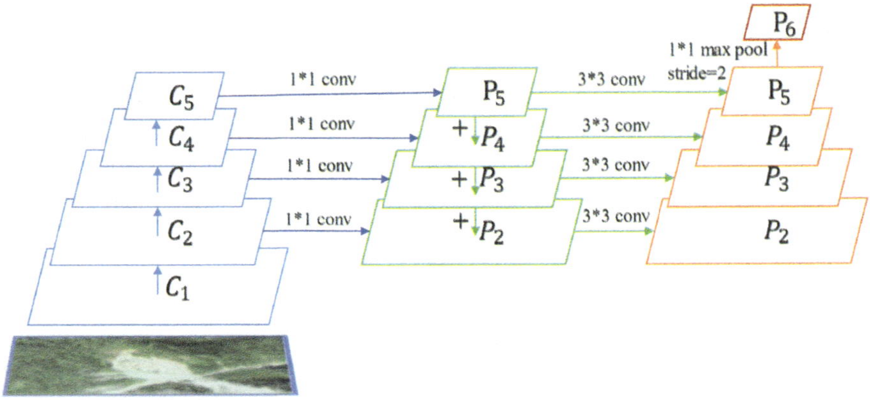


**Fig. 5.9** Network structure of FPN, modified from Lin et al., (2017a, 2017b)

level is lower, so upsampling is needed to keep the resolution consistent with that of the previous stage. Finally, feature maps of adjacent stages with unified resolution and channel amount are added for feature fusion.

As shown in Fig. 5.9, the bottom-up process is on the left and the top-down process is in the middle. Bottom-up is also known as the forward process of convolutional network. After each pooling, the feature map of the next stage is obtained. It can be seen that $C_1$ is not used, because the high resolution of $C_1$ convolutional network will lead to high video memory requirement in the training process. In the top-down process, the horizontal connection strategy is used to obtain four outputs $P_2$, $P_3$, $P_4$, $P_5$. The $3 \times 3$ convolutions are used for these four outputs to reduce the aliasing effect of upsampling. Finally, the $1 \times 1$ maximum pooling with stride size of 2 is used for $P_5$ to downsample to get $P_6$.

FPN is a pioneer in feature fusion. By combining deep semantic information with shallow texture information, it gives the feature map of each layer with stronger ability to capture semantic information. As it only provides a feature fusion mechanism, few parameters are introduced, and the detection ability of multi-scale targets is improved without increasing much computational amount.

### 5.3.3 *Faster R-CNN Based on Graph Convolution and Feature Pyramid*

Target detection algorithm framework usually consists of four parts: backbone network + regional suggestion network + RoI pooling layer + fully connection classifier. In order to improve the network's ability to detect small targets, the feature pyramid is added into the Faster R-CNN as neck part, as shown in Fig. 5.10.

The model firstly inputs the image derived from landslide dataset into the backbone network for bottom-up process to extract features. ResNet50 is used as the backbone

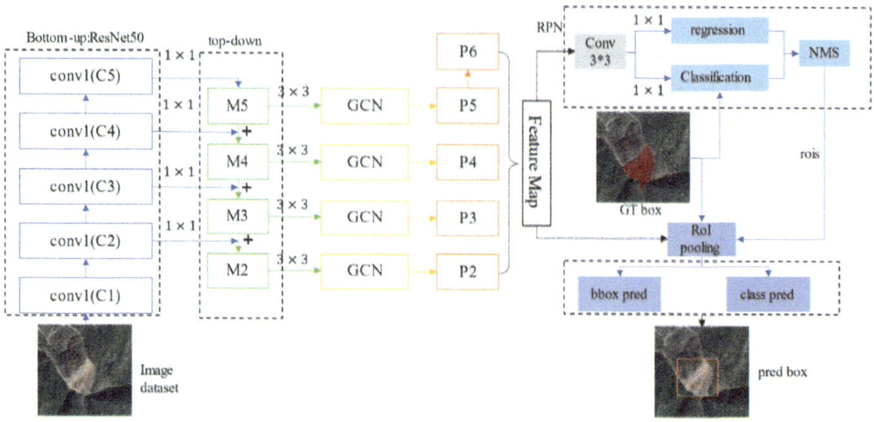


**Fig. 5.10** The Faster R-CNN network structure based on graph convolution and feature pyramid

network. Then top-down process is based on the FPN feature fusion algorithm. Fusion results are afterwards processed by GCN module. Since the resolutions of levels are different, four GCN modules are used in total, which contains three graph convolution layers. After the feature fusion, five feature maps in different scales are generated. These feature maps are input into the RPN network to extract candidate boxes, and 9 anchors with different scales and proportions are generated for each feature point according to the anchor box mechanism introduced previously. Information of GT box labeling is also needed to input into the RPN network. Based on GT box information, positive and negative samples of anchor are labelled. And 256 anchors are selected to classify the foreground and background, and perform rough position coordinate regression. Meanwhile, RPN also needs to provide RoIs for the RoI pooling layer as training samples. By calculating the probability for each anchor belonging to positive sample and by using the non-maximum suppression algorithm (NMS), RoIs are selected and delivered to the RoI pooling layer. RoI pooling layer normalizes all candidate regions to a unified scale, then classification and position regression are performed in the two fully connected layers.

In the Faster R-CNN, classification and position regression share features, which enables them to be trained at the same time. Therefore, a multi-task loss function composed of classification error and regression error is constructed:

$$L(\{p_i\}, \{t_i\}) = \frac{1}{N_{cls}} \sum_i L_{cls}(p_i, p_i^*) + \lambda \frac{1}{N_{reg}} \sum_i p_i^* L_{reg}(t_i, t_i^*) \tag{5.21}$$

where $N_{reg}$ is the size of the input feature map, $N_{cls}$ is the number of anchors in a batch, $i$ is the index of anchor, $\lambda$ is the balanced proportion parameter of classification and regression, usually equals to 10. $L_{cls}(p_i, p_i^*)$ is the classification loss function to determine if there is a landslide. $p_i$ represents the probability that an anchor contains the target, and $p_i^*$ is the binary label assigned to anchor. $L_{reg}(t_i, t_i^*)$ is the loss of bounding box regression. $t_i$ represents the 4 parameter coordinates of the prediction box, and $t_i^*$ represents the real coordinates of the mark box containing landslide.

In Eq. (5.21), classification loss is calculated by cross entropy, and $L_{cls}(p_i, p_i^*)$ calculation method is as follows:

$$L_{cls}(p_i, p_i^*) = -\log[p_i^* p_i + (1 - p_i^*)(1 - p_i)] \tag{5.22}$$

The regression error $L_{reg}(t_i, t_i^*)$ is calculated as follows:

$$L_{reg}(t_i, t_i^*) = R(t_i - t_i^*) \tag{5.23}$$

where parameter R is the smooth function, and the calculation is as follows:

$$\text{smooth}_{L_1}(x) = \begin{cases} 0.5x^2, & |x| < 1 \\ |x| - 0.5, & \text{otherwise} \end{cases} \tag{5.24}$$

## 5.4 Experiments

### *5.4.1 Experiments Settings*

The programming framework of the computer environment and relevant codes is shown in Table 5.1.

MMDetection is an open source project that can customize model and data for target detection taks, and support most of the target detection framework. Comparing to other open-source projects, it has higher performance in precision, with faster training speed and lower memory requirement. Faster R-CNN in this study is implemented by using this toolkit.

**Table 5.1** Environment configuration

| Experimental configuration | Version and model |
|---|---|
| System environment | Windows10 |
| Memory size | 16G |
| The CPU model | Intel (R) Core (TM) I5-8300H |
| GPU | NVIDIA GeForce GTX 1050 |
| Programming language | Python3.9.7 |
| Deep learning library | PyTorch1.7.1 + cu110 |
| | TorchVision0.8.2 + cu110 |
| Target detection open source framework | MMDetection2.20.0 |

The network hyperparameters are set as follow: The batch size is set to 2. The training times are 30 epochs, and each epoch was iterated for 180 times, with a total of 5400 iterations. The original learning rate is set to 0.02 for the stochastic gradient descent (SGD) optimization algorithm, with the momentum of 0.9, and the weight attenuation coefficient of 0.0001.

### 5.4.2 Dataset Source

Image data used in this research is mainly collected from Gaofen-2 panchromatic and multispectral satellites during 2017–2019. And the data from Gaofen-3 and Gaofen-7 are used as auxiliary data to build the datasets. Geometric correction is performed by using polynomial correction and tonal adjustment is achieved by histogram specification method. The basic information of landslides is obtained by the investigation in 221 villages, 11 townships within 8994 $km^2$ of Wangqing County, Yanbian Chaoxian (Korean) Nationality Autonomous, China. The spatial distribution of landslides collected in the dataset is shown in Fig. 5.11.

ArcGIS software is used for the visual interpretation of the collected image data, in combination of the interpretation symbols of unstable landslides and previous materials. Coordinate batch cropping is adopted to subset the images to a unified size of $600 \times 600$. According to the location information of landslide, Labelme was used to label the clipped image to generate landslide dataset, which is then normalized to the standard format of COCO dataset with unified image file naming and format, as shown in Fig. 5.12.

The COCO dataset was a single-category landslide dataset consists of 520 images in 1 m spatial resolution. The dataset contains multiple types of landslide remote sensing image targets: large scale landslide target as shown in Fig. 5.12a, small scale landslide target Fig. 5.12b, significant landslide target Fig. 5.12c and fuzzy landslide target Fig. 5.12d. Train_test_split toolkit in Sklearn module is used to gain the train set with 70% images (364) and test set with 30% images (156) (Fig. 5.13).

### 5.4.3 Experiment Content

#### Evaluation Methods

In this research, precision (P), recall (R), accuracy (Acc), average accuracy (AP) and PR curve are used to evaluate the model performance basically. Precision refers to the proportion of real landslide in all positive sample images identified. Recall refers to the proportion of all true landslides in the test set that are correctly identified as positive sample landslides. The higher the recall value is, the more images

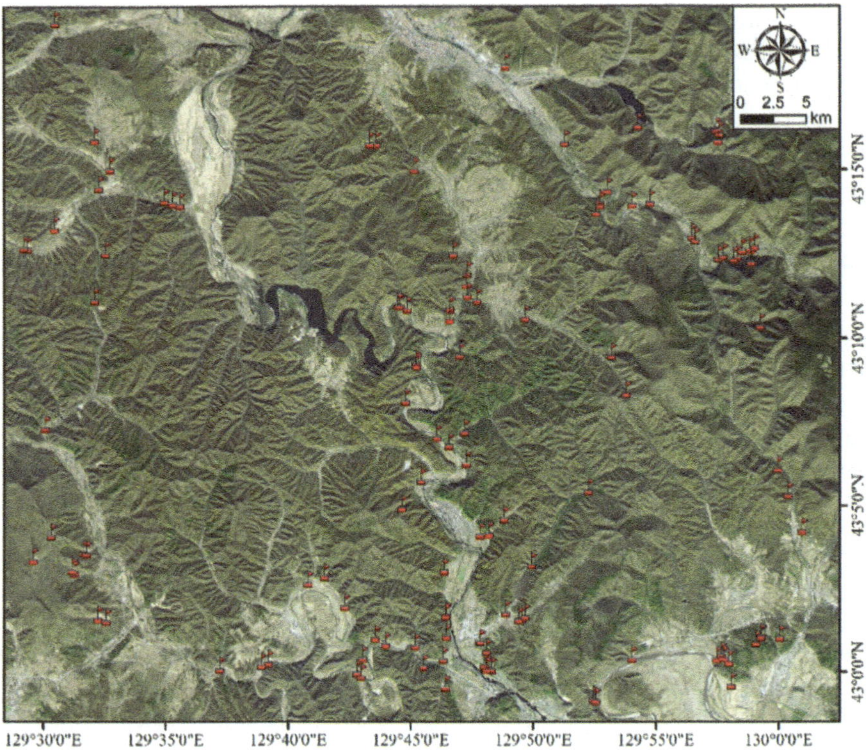


**Fig. 5.11** Locations of the landslides in Wangqing County, Yanbian Chaoxian (Korean) Nationality Autonomous, China

are detected correctly. Since the difference of confidence between recall and precision can affect each other, AP is introduced for a comprehensive evaluation of performance. Equations for these metrics are:

$$P = \frac{TP}{TP + FP} \times 100\% \tag{5.25}$$

$$R = \frac{TP}{TP + FN} \times 100\% \tag{5.26}$$

$$Acc = \frac{TP + TN}{TP + TN + FP + FN} \times 100\% \tag{5.27}$$

The definition of parameters is shown in Table 5.2.

The mAP is the average value of all target categories. Since this experiment is a single type of landslide target detection, mAP is the average precision of landslide. The mAP0.5 means that the threshold of IoU is set as 0.5. In the experiment, the

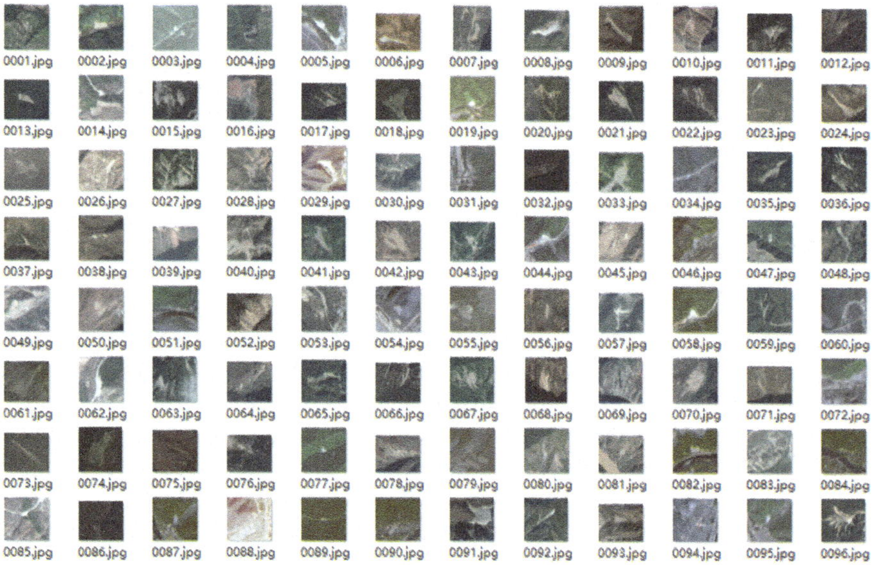


**Fig. 5.12** Landslide dataset

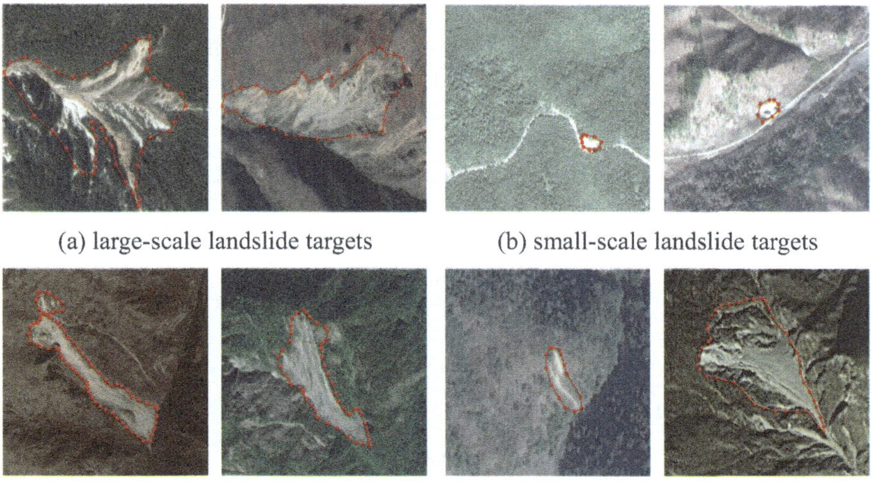

(a) large-scale landslide targets (b) small-scale landslide targets

(c) significant landslide targets (d) fuzzy landslide targets

**Fig. 5.13** Remote sensing image targets of landslides in different forms

**Table 5.2** Parameter definitions

| Category | Postive sample | Negative sample |
|---|---|---|
| True | TP | TN |
| False | FP | FN |

threshold of mAP is set as [0.5:0.05:0.95]. Ten mAP values are extracted from 0.5 to 0.95 with a 0.05 interval, which are then averaged to obtain mAP.

PR curve is drawn with recall as the horizontal axis and precision as the vertical axis. With the increase amount of detections, the curve gradually tends to be stable. In this case, the area enclosed by the curve and the horizontal axis is the AP value. When there are multiple types of targets, there will be multiple AP values, which can be averaged to mAP. In this experiment, landslide is detection target, so AP value equals to mAP.

**Training Results**

The performance of original Faster R-CNN model with the Faster R-CNN model contains GCN is compared in this section. The size of the image is limited to 128 × 128 due to the video memory of the machine.

In Fig. 5.14, the losses of classification and coordinate regression are calculated simultaneously. The smaller the Loss value is, the closer the detection result of the model will be to the true detection box, and the classification will be more accurate. Since the loss difference between the two model is not significant, the loss value is plotted logarithmically to make the difference be more obvious. After 25 epochs (about 4000 iterations), the loss value of the Faster R-CNN model with GCN module is lower than that of the original one. Loss value also keeps decreasing and tends to be stable.

Figure 5.15 shows the variation trend of the accuracy with iterations. In early epochs, the accuracy of Faster R-CNN added with GCN module is lower than that of the original Faster R-CNN. However, with the increase of iteration numbers, GCN

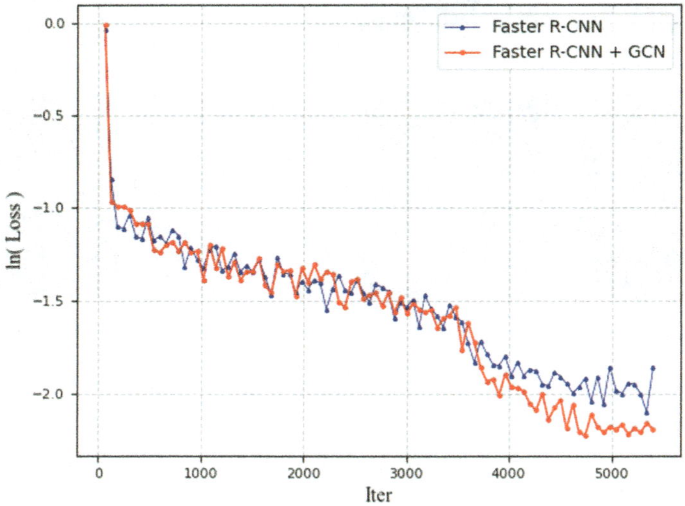


**Fig. 5.14** Loss comparison diagram of two Faster R-CNN models

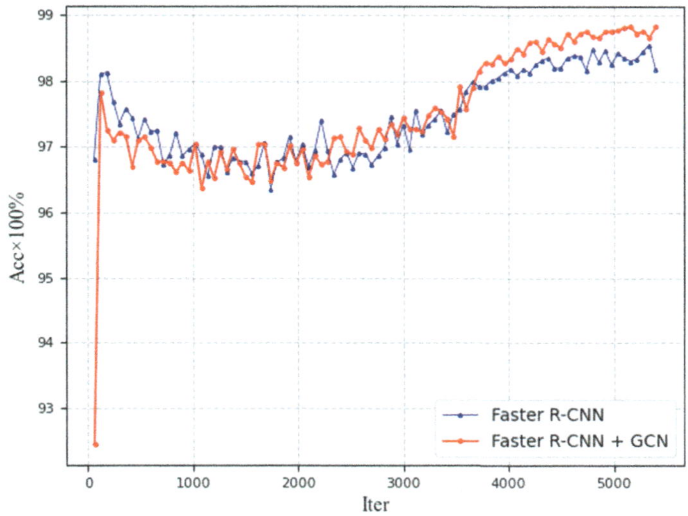


**Fig. 5.15** Accuracy comparison diagram of two faster R-CNN models

continuously optimizes the model parameters. The accuracy keeps increasing and exceeds that of the original Faster R-CNN model.

**Testing Results**

In the 30 epochs of model training, the model parameters of the 25th epoch with lowest loss value and higher precision were selected to build the optimal model and evaluate the performance. The results are shown in Table 5.3. Precision and recall of the Faster R-CNN model are both lower than those of the Faster R-CNN model added with GCN, but them metric differences are insignificantly. Overall, the performance of landslide detection model with GCN module is slightly better than that of the original model. The mAP values are calculated from the PR curve in Fig. 5.16. The threshold value in this experiment is equal to 0.05.

The landslide detection results of the test set by using the Faster R-CNN with GCN module are indicated in Fig. 5.17. And the detection results are also compared with the true landslide labeling box.

**Table 5.3** Test results

| Algorithm | mAP (%) | Precision (%) | Recall (%) |
|---|---|---|---|
| Faster R-CNN | 40.4 | 60.1 | 40.2 |
| Faster R-CNN + GCN | 41.3 | 60.9 | 41.9 |

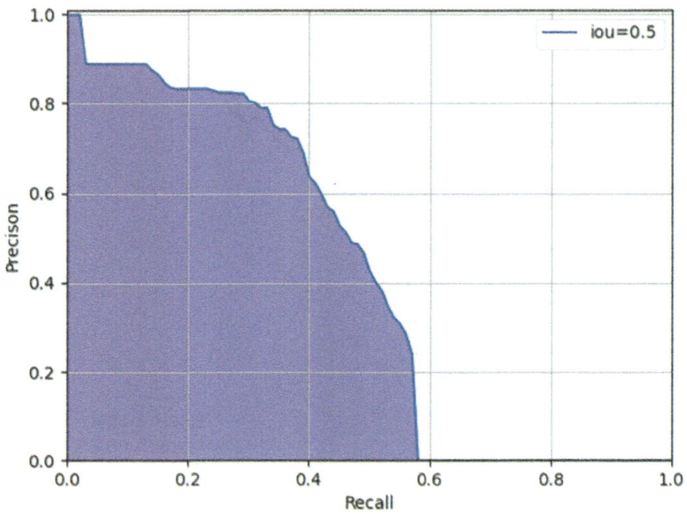


**Fig. 5.16** PR curve

## 5.5 Summary

Receptive field of CNN network is usually confined around the convolution kernel. To exchange information over a long distance, multiple convolutions or pool operations are required to reduce the scale at the cost of information loss. GCN can aggregate the features of distant neurons, which allow it to have a larger receptive field. This chapter mainly studies the fusion of GCN and Faster R-CNN algorithm to improve the effect of landslide detection. The main contents are as follows:

(1) The research background and the current status of landslide target detection are reviewed. Popular target detection frameworks based on CNN are introduced.
(2) Graph convolution method is used in combination with Faster R-CNN. Considering different connection modes, a graph convolution layer is designed which can automatically learn adjacency matrix, so that the model can learn different connection mode according to different data, which makes the model more robust.
(3) The proposed target detection framework is compared with the original model without GCN module. All of the precision, recall and mAP of proposed framework are about 1% higher than that of the original model.

There are still many limitations of the model:

(1) Due to limited computer performance, the hyperparameters of the model are not fully debugged.
(2) Only 520 landslide images are used in this experiment, which is relatively small for the deep learning model.

Some potential questions worth further exploration:

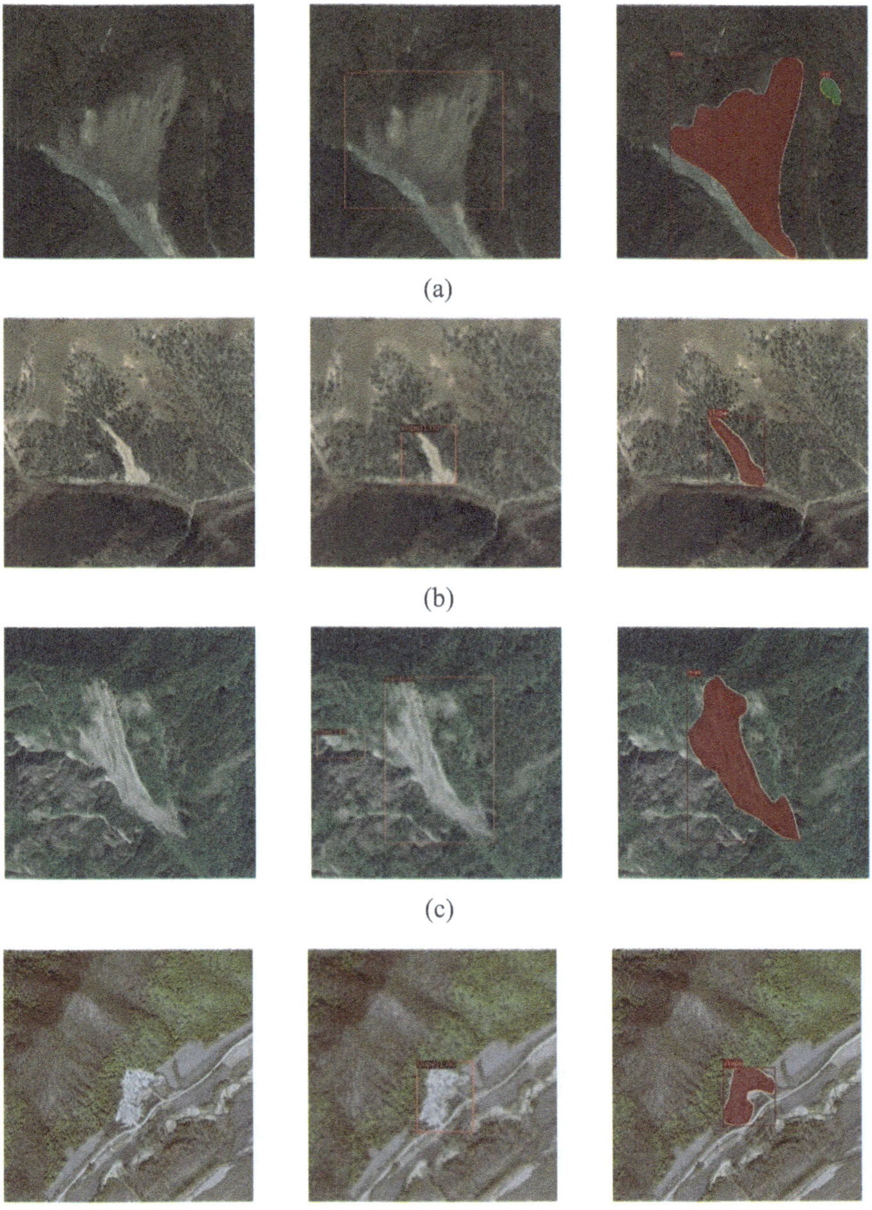

(a) (b) (c) (d)

**Fig. 5.17** Comparison of landslide detection results and true landslide labeling. The input images are on the left, the images with detection results of the model are in the middle, and images with the true landslide labeling box are on the right

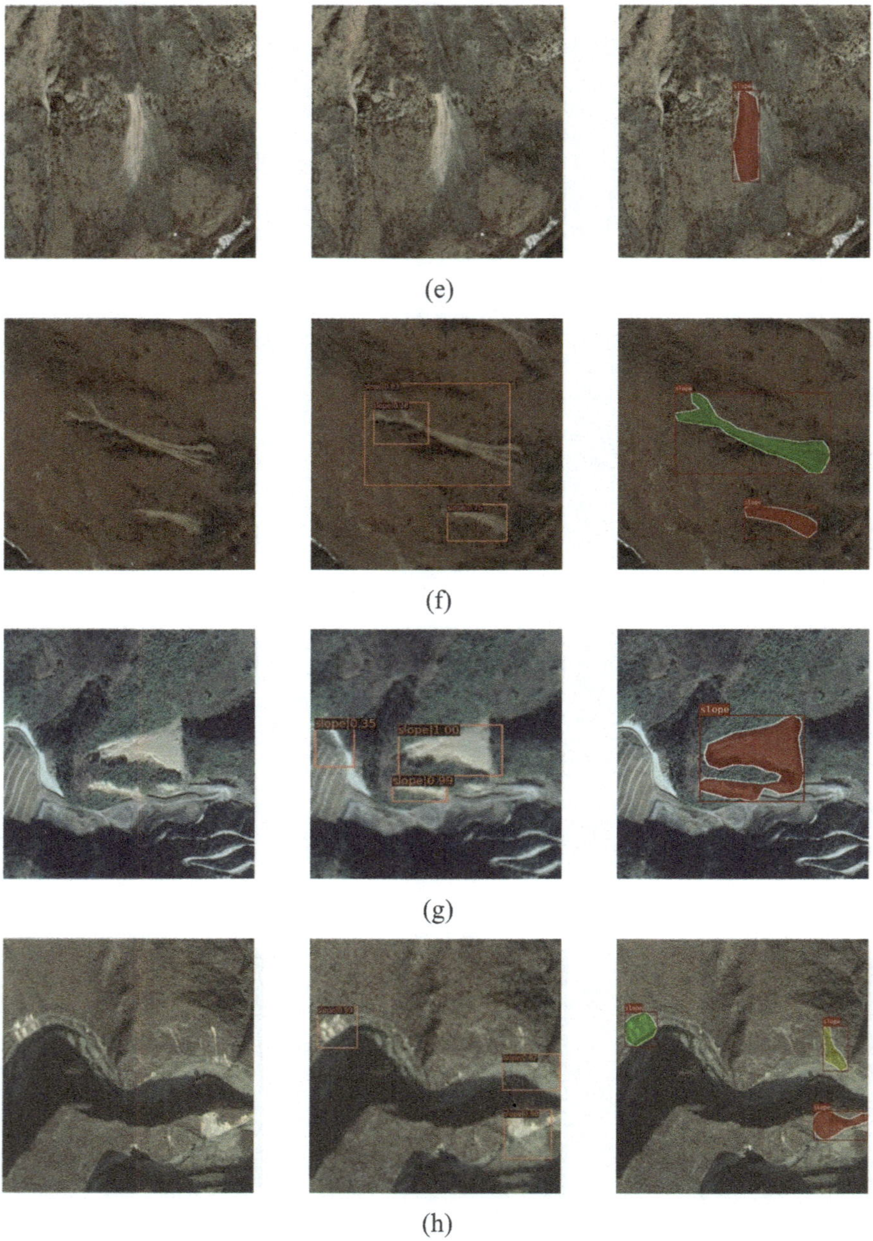


**Fig. 5.17** (continued)

(1) The weights of parameter matrix of GCN are shared, which reduces the number of parameters, but also leads to the same parameter number of each neuron. Therefore, semantic graph convolution has been proposed to solve this problem, which an be applied to landslide detection in the future.
(2) As the environment of landslide is complex and diverse, well pre-process of landslide images will be helpful to improve the detection accuracy.

## Codes

### [Python3.9] Faster_rcnn_r50_fpn_2x_coco

```
model = dict(
   type = 'FasterRCNN',
   backbone = dict(
      type = 'ResNet',
      depth = 50,
      num_stages = 4,
      out_indices = (0, 1, 2, 3),
      frozen_stages = 1,
      norm_cfg = dict(type = 'BN', requires_grad = True),
      norm_eval = True,
      style = 'pytorch',
       init_cfg = dict(type = 'Pretrained', checkpoint = 'torchvi-
sion://resnet50')),
   neck = dict(
      type = 'FPN',
      in_channels = [256, 512, 1024, 2048],
      out_channels = 256,
      num_outs = 5),
   rpn_head = dict(
      type = 'RPNHead',
      in_channels = 256,
      feat_channels = 256,
      anchor_generator = dict(
         type = 'AnchorGenerator',
         scales = [8],
         ratios = [0.5, 1.0, 2.0],
         strides = [4, 8, 16, 32, 64]),
      bbox_coder = dict(
         type = 'DeltaXYWHBBoxCoder',
         target_means = [0.0, 0.0, 0.0, 0.0],
         target_stds = [1.0, 1.0, 1.0, 1.0]),
      loss_cls = dict(
        type = 'CrossEntropyLoss', use_sigmoid = True, loss_weight
= 1.0),
      loss_bbox = dict(type = 'L1Loss', loss_weight = 1.0)),
   roi_head = dict(
      type = 'StandardRoIHead',
      bbox_roi_extractor = dict(
         type = 'SingleRoIExtractor',
      roi_layer = dict(type = 'RoIAlign', output_size = 7, sampling_
ratio = 0),
```

```
            out_channels = 256,
            featmap_strides = [4, 8, 16, 32]),
        bbox_head = dict(
            type = 'Shared2FCBBoxHead',
            in_channels = 256,
            fc_out_channels = 1024,
            roi_feat_size = 7,
            num_classes = 1,
            bbox_coder = dict(
                type = 'DeltaXYWHBBoxCoder',
                target_means = [0.0, 0.0, 0.0, 0.0],
                target_stds = [0.1, 0.1, 0.2, 0.2]),
            reg_class_agnostic = False,
            loss_cls = dict(
                  type = 'CrossEntropyLoss', use_sigmoid = False, loss_
weight = 1.0),
            loss_bbox = dict(type = 'L1Loss', loss_weight = 1.0))),
    train_cfg = dict(
        rpn = dict(
            assigner = dict(
                type = 'MaxIoUAssigner',
                pos_iou_thr = 0.7,
                neg_iou_thr = 0.3,
                min_pos_iou = 0.3,
                match_low_quality = True,
                ignore_iof_thr = -1),
            sampler = dict(
                type = 'RandomSampler',
                num = 256,
                pos_fraction = 0.5,
                neg_pos_ub = -1,
                add_gt_as_proposals = False),
            allowed_border = -1,
            pos_weight = -1,
            debug = False),
        rpn_proposal = dict(
            nms_pre = 2000,
            max_per_img = 1000,
            nms = dict(type = 'nms', iou_threshold = 0.7),
            min_bbox_size = 0),
        rcnn = dict(
            assigner = dict(
                type = 'MaxIoUAssigner',
                pos_iou_thr = 0.5,
                neg_iou_thr = 0.5,
                min_pos_iou = 0.5,
                match_low_quality = False,
                ignore_iof_thr = -1),
            sampler = dict(
                type = 'RandomSampler',
                num = 512,
                pos_fraction = 0.25,
                neg_pos_ub = -1,
                add_gt_as_proposals = True),
```

```
            pos_weight = -1,
            debug = False)),
    test_cfg = dict(
        rpn = dict(
            nms_pre = 1000,
            max_per_img = 1000,
            nms = dict(type = 'nms', iou_threshold = 0.7),
            min_bbox_size = 0),
        rcnn = dict(
            score_thr = 0.05,
            nms = dict(type = 'nms', iou_threshold = 0.5),
            max_per_img = 100)))
dataset_type = 'CocoDataset'
data_root = 'data/coco/'
img_norm_cfg = dict(
   mean = [123.675, 116.28, 103.53], std = [58.395, 57.12, 57.375],
to_rgb = True)
train_pipeline = [
    dict(type = 'LoadImageFromFile'),
    dict(type = 'LoadAnnotations', with_bbox = True),
   dict(type = 'Resize', img_scale = (128, 128), keep_ratio = True),
    dict(type = 'RandomFlip', flip_ratio = 0.5),
    dict(
        type = 'Normalize',
        mean = [123.675, 116.28, 103.53],
        std = [58.395, 57.12, 57.375],
        to_rgb = True),
    dict(type = 'Pad', size_divisor = 32),
    dict(type = 'DefaultFormatBundle'),
   dict(type = 'Collect', keys = ['img', 'gt_bboxes', 'gt_labels'])
]
test_pipeline = [
    dict(type = 'LoadImageFromFile'),
    dict(
        type = 'MultiScaleFlipAug',
        img_scale = (128, 128),
        flip = False,
        transforms = [
            dict(type = 'Resize', keep_ratio = True),
            dict(type = 'RandomFlip'),
            dict(
                type = 'Normalize',
                mean = [123.675, 116.28, 103.53],
                std = [58.395, 57.12, 57.375],
                to_rgb = True),
            dict(type = 'Pad', size_divisor = 32),
            dict(type = 'ImageToTensor', keys = ['img']),
            dict(type = 'Collect', keys = ['img'])
        ])
]
data = dict(
    samples_per_gpu = 2,
    workers_per_gpu = 2,
    train = dict(
```

```
        type = 'CocoDataset',
       ann_file = 'data/coco/annotations/instances_train2017.json',
        img_prefix = 'data/coco/train2017/',
        pipeline = [
            dict(type = 'LoadImageFromFile'),
            dict(type = 'LoadAnnotations', with_bbox = True),
            dict(type = 'Resize', img_scale = (128, 128), keep_ratio =
True),
            dict(type = 'RandomFlip', flip_ratio = 0.5),
            dict(
                type = 'Normalize',
                mean = [123.675, 116.28, 103.53],
                std = [58.395, 57.12, 57.375],
                to_rgb = True),
            dict(type = 'Pad', size_divisor = 32),
            dict(type = 'DefaultFormatBundle'),
              dict(type = 'Collect', keys = ['img', 'gt_bboxes', 'gt_
labels'])
        ]),
    val = dict(
        type = 'CocoDataset',
        ann_file = 'data/coco/annotations/instances_val2017.json',
        img_prefix = 'data/coco/val2017/',
        pipeline = [
            dict(type = 'LoadImageFromFile'),
            dict(
                type = 'MultiScaleFlipAug',
                img_scale = (128, 128),
                flip = False,
                transforms = [
                    dict(type = 'Resize', keep_ratio = True),
                    dict(type = 'RandomFlip'),
                    dict(
                        type = 'Normalize',
                        mean = [123.675, 116.28, 103.53],
                        std = [58.395, 57.12, 57.375],
                        to_rgb = True),
                    dict(type = 'Pad', size_divisor = 32),
                    dict(type = 'ImageToTensor', keys = ['img']),
                    dict(type = 'Collect', keys = ['img'])
                ])
        ]),
    test = dict(
        type = 'CocoDataset',
        ann_file = 'data/coco/annotations/instances_val2017.json',
        img_prefix = 'data/coco/val2017/',
        pipeline = [
            dict(type = 'LoadImageFromFile'),
            dict(
                type = 'MultiScaleFlipAug',
                img_scale = (128, 128),
                flip = False,
                transforms = [
                    dict(type = 'Resize', keep_ratio = True),
```

```
                dict(type = 'RandomFlip'),
                dict(
                    type = 'Normalize',
                    mean = [123.675, 116.28, 103.53],
                    std = [58.395, 57.12, 57.375],
                    to_rgb = True),
                dict(type = 'Pad', size_divisor = 32),
                dict(type = 'ImageToTensor', keys = ['img']),
                dict(type = 'Collect', keys = ['img'])
            ])
        ]))
evaluation = dict(interval = 1, metric = 'bbox')
optimizer = dict(type = 'SGD', lr = 0.02, momentum = 0.9, weight_
decay = 0.0001)
optimizer_config = dict(grad_clip = None)
lr_config = dict(
    policy = 'step',
    warmup = 'linear',
    warmup_iters = 500,
    warmup_ratio = 0.001,
    step = [20, 25])
runner = dict(type = 'EpochBasedRunner', max_epochs = 30)
checkpoint_config = dict(interval = 1)
log_config = dict(interval = 60, hooks = [dict(type = 'TextLogger-
Hook')])
custom_hooks = [dict(type = 'NumClassCheckHook')]
dist_params = dict(backend = 'nccl')
log_level = 'INFO'
load_from = None
resume_from = None
workflow = [('train', 1)]
work_dir = './work_dirs\faster_rcnn_r50_fpn_2x_coco'
auto_resume = False
gpu_ids = range(0, 1)
```

## References


Behling, R., Roessner, S., Kaufmann, H., & Kleinschmit, B. (2014). Automated spatiotemporal landslide mapping over large areas using rapideye time series data. *Remote Sensing, 6*(9), 8026–8055.

Bruna, J., Zaremba, W., Szlam, A., & LeCun, Y. (2013). *Spectral networks and locally connected networks on graphs*. arXiv preprint, arXiv:1312.6203.

Chen, K., Zhu, Z., Deng, X., Ma, C., & Wang, H. (2021). Deep learning for multi-scale object detection: A survey. *Journal of Software, 32*(4), 1201–1227.

Chen, W., Pourghasemi, H. R., & Zhao, Z. (2017). A GIS-based comparative study of Dempster-Shafer, logistic regression and artificial neural network models for landslide susceptibility mapping. *Geocarto International, 32*(4), 367–385.

Chen, Y., Sun, H., Zhao, X., Tang, H., & Zhou, T. (2011). Remote sensing detection analysis of valley landslide in earthquake disaster area: A case study of landslides in Beichuan County. *Journal of Natural Disasters, 20*(1), 97–104.

Girshick, R., Donahue, J., Darrell, T., & Malik, J. (2014). Rich feature hierarchies for accurate object detection and semantic segmentation. In *Proceedings of the IEEE conference on computer vision and pattern recognition*, pp. 580–587.

Girshick, R. (2015). Fast r-cnn. In *Proceedings of the IEEE international conference on computer vision*, pp. 1440–1448.

Guo, J., Li, Y., Li, Z., Liu, K., & Zhang, S. (2016). An automatic interpretation model for mountains landslide disaster of high-resolution remote sensing images based on transfer learning. *Journal of Geomatics Science and Technology, 33*(5), 496–501.

He, K., Zhang, X., Ren, S., & Sun, J. (2016). Deep residual learning for image recognition. In *Proceedings of the IEEE conference on computer vision and pattern recognition*, pp. 770–778.

He, K., Zhang, X., Ren, S., & Sun, J. (2015). Spatial pyramid pooling in deep convolutional networks for visual recognition. *IEEE Transactions on Pattern Analysis and Machine Intelligence, 37*(9), 1904–1916.

Huang, R. (2009). Mechanism and geomechanical modes of landslide hazards triggered by Wenchuan 8.0 Earthquake. *Chinese Journal of Rock Mechanics and Engineering, 28*(6), 1239–1249.

Huang, T., Bai, X., Zhuang, Q., & Xu, J. (2018). Research on landslides extraction based on the wenchuan earthquake in GF-1 remote sensing image. *Bulletin of Surveying and Mapping, 128*, 67–71.

Ju, Y., Xu, Q., Jin, S., Li, W., Dong, X., & Guo, Q. (2020). Automatic object detection of loess landslide based on deep learning. *Geomatics and Information Science of Wuhan University, 45*(11), 1747–1755.

Li, L., Wu, Y., Miao, F., Zhang, L., & Xue, Y. (2019). Landslide displacement interval prediction based on different Bootstrap methods and KELM-BPNN model. *Chinese Journal of Rock Mechanics and Engineering, 38*(5), 912–926.

Lin, T.-Y., Dollár, P., Girshick, R., He, K., Hariharan, B., & Belongie, S. (2017a). Feature pyramid networks for object detection. In *Proceedings of the IEEE conference on computer vision and pattern recognition*, pp. 2117–2125.

Lin, T.-Y., Goyal, P., Girshick, R., He, K., & Dollár, P. (2017b). Focal loss for dense object detection. In *Proceedings of the IEEE international conference on computer vision*, pp. 2980–2988.

Liu, W., Anguelov, D., Erhan, D., Szegedy, C., Reed, S., Fu, C.-Y., & Berg, A. C. (2016). Ssd: Single shot multibox detector. In *European conference on computer vision*, pp. 21–37.

Pham, B. T., Bui, D. T., Prakash, I., & Dholakia, M. B. (2017). Hybrid integration of multilayer perceptron neural networks and machine learning ensembles for landslide susceptibility assessment at Himalayan area (India) using GIS. *CATENA, 149*, 52–63.

Redmon, J., Divvala, S., Girshick, R., & Farhadi, A. (2016). You only look once: Unified, real-time object detection. In *Proceedings of the IEEE conference on computer vision and pattern recognition*, pp. 779–788.

Ren, S., He, K., Girshick, R., & Sun, J. (2015). Faster r-cnn: Towards real-time object detection with region proposal networks. *Advances in Neural Information Processing Systems*.

Shi, Y., Shi, L., Niu, M., & Zhao, K. (2022). Multi-task automatic identification of loess landslide based on one-stage instance segmentation network. *Bulletin of Surveying and Mapping, 4*, 26–31.

Szegedy, C., Toshev, A., & Erhan, D. (2013). Deep neural networks for object detection. *Advances in Neural Information Processing Systems*.

Zhang, Q., & Zhao, C. (2017). Semiautomatic object-oriented loose landslide recognition based on high resolution remote sensing images in Heifangtai, Gansu. *Journal of Catastrophology, 32*(3), 210–215.

Zhou, F., Jin, L., & Dong, J. (2017). Review of convolutional neural network. *Chinese Journal of Computers, 40*(6), 1230–1251.

# Chapter 6
# Deep Learning Based Landslide Susceptibility Assessment

**Abstract** Susceptibility is an important indicator in the prevention of geodisasters. China has a diverse geological environment and contains many potential geological hazards that are difficult for management. This chapter proposed a geodisaster susceptibility assessing framework based on graph convolutional network (GCN) to assess the landslide susceptibility in Yanbian Korean Autonomous Prefecture, China. Firstly, features related to landslide are selected as model inputs using Pearson correlation coefficient method and are converted to two-dimensional forms to construct the landslide dataset. Then a GCN landslide susceptibility assessing model is established and trained and tested on the prepared dataset. Finally, the predicted susceptibility indices are graded using the natural breakpoint method to generate the landslide susceptibility zoning map. The results indicate that the model used in this paper has good generalization ability, higher prediction accuracy comparing with other deep learning models. The framework can help the regional stakeholders to improve the ability to prevent and control landslide.

**Keywords** Landslide · Susceptibility assessment · Graph neural network · Convolutional neural network

## 6.1 Introduction

### *6.1.1 Background and Significance*

Geodisasters involve geological processes inside the earth like earthquake, volcanic activity and gas emission, and geological processes on the surface of the earth like landslide, debris flow, and avalanche. Both natural and anthropogenic activities can lead to geodisasters. Usually, these disasters can destroy the local infrastructures and threaten the human life, which seriously restrict the sustainable development of regional economy.

China has large land area with diverse geological environments which leads to the high numbers of potential geodisasters. Only a small part of the geodisasters distribute concentratedly, while most of them distribute separately (Zhu, 2020). This

 
W. Chen et al., *Intelligent Interpretation for Geological Disasters*,
https://doi.org/10.1007/978-981-99-5822-1_6

causes great difficulties for the disaster prevention and management. According to the Ministry of Natural Resources, it was reported that 4772 geological disasters occurred in 2021, resulting in 80 deaths and 11 missing cases, The direct economic losses was about 3.2 billion RMB. Regarding the disaster type, the reported geodisasters include 2335 landslides, 1746 collapses, 374 debris flows, 285 ground collapses, 21 ground fissures, and 11 ground subsidence. As for disaster intensity, they include 35 super-large geological disasters, 27 large geological disasters, 328 medium geological disasters, and 4382 micro geological disasters (Ministry of Natural Resources of the People's Republic of China, 2022a, 2022b). Accurate prediction of geodisasters can significantly reduce the influence of geodisasters on human society. China completed the predictions of 905 geodisasters in 2021, avoiding the direct economic losses of 1.35 billion RMB. Compared with the average performance of same time period in the previous 5 years, the number of geological disasters, and deaths and missing cases decreased by 30.3 and 63.2%, respectively. The direct economic losses remained the same (Ministry of Natural Resources of the People's Republic of China, 2022a, 2022b).

The geodisaster susceptibility zoning makes it possible to prevent geodisasters, determine control measures and develop geo-environmental resources reasonably. Zoning based on the geological environment can effectively indicate the history and possibility of future occurrence of geodisasters (Yang, 2016). Since the beginning of this century, many researchers have conducted studies on geodisasters susceptibility, and some of the findings provide theoretical and methodological references for this study. GIS technology is a reliable source for the basic data of geological disasters susceptibility. It can be used to select the related factors from the geological environment factors. Based on this, the evaluation model of geodisasters susceptibility is proposed using qualitative and quantitative methods. With further analysis, the geodisasters susceptibility zoning map can be finally gained.

The susceptibility assessment of geodisasters is a valuable reference for geological stability prediction, which is of important research significance and is a realistic demand for urban construction and land development. With the wide promotion of deep learning (DL) technology, the application of DL models in geological disasters susceptibility zoning has achieved good performance in recent years. This research proposes a robust DL-based geodisasters susceptibility evaluation model to reduce the dependence on prior knowledge in zoning and improve the accuracy of susceptibility assessment. The model can help the regional stakeholders to improve the ability to prevent and control geodisasters.

### *6.1.2 Research Overview*

#### Geodisasters Susceptibility Zoning with Traditional Methods

Research on the susceptibility evaluation of geodisasters started from the 1970s. With the rapid development of GIS technology after the 1990s, researchers have created a variety of geodisasters prediction models based on the GIS-related technologies. Commonly used models include analytic hierarchy process (AHP), fuzzy and gray clustering method, information model evaluation method, and empirical weighting method.

Zhang et al. (1993) established a systematic comprehensive evaluation model of geodisasters, in which the comprehensive gray scale was used as an evaluation metric to determine the degree of destroy. Comprehensive disaster indices were calculated by the AHP method to analyze the distribution of geodisasters in China. Wang et al. (2012) established a geodisaster susceptibility evaluation system with an ensemble approach to determine the contribution of the indices to the occurrence of geodisasters. A fuzzy judgment matrix was used for the second-level fuzzy evaluation to achieve the susceptibility zoning. Sun et al. (2008) used the information quantity model to evaluate the study area with series of indicators were selected as the model inputs for the susceptibility evaluation. Mallick et al. (2018) proposed a method called fuzzy analytic hierarchical process (FAHP). The fuzzy membership function was first used to determine the criteria weights of susceptibility. Based on GIS-based technologies and statistical modeling, landslide susceptibility areas were evaluated by FAHP and feature vector method. Finally weighted linear combination method was used to generate landslide susceptibility zoning map. Grozavu (2021) firstly divided Romania into five climatic-geographic zones considering landslide susceptibility factors of slope, lithology, land cover, etc. Then the factor weights were obtained using AHP based on frequency ratios and expert knowledge. Multiplication of factor weights and category weights were summed to calculate the susceptibility indices and generate the zoning maps.

These traditional methods abovementioned highly depend on the prior knowledge, such as subjective determination of weights by experts to evaluate causative factors, which may lead to different results and cause the higher uncertainty.

**Geodisasters Susceptibility Zoning with Machine Learning**

Machine learning models depend on a supervised feedback training mechanism that allows the models to be trained gradually until they are close to the real situations. Machine learning models mainly include logistic regression model, artificial neural network (ANN) (Oh & Lee, 2017), support vector machine (SVM) (Hong et al., 2016), decision tree (DT) (Chen et al., 2017), Bayesian networks, random forest (RF) (Liang et al., 2020), etc. These models are widely used and work well in the engineering practice of geodisaster susceptibility assessment, and have been proved to be effective ways for geodisaster prediction (Hu, 2020).

Tang and Ma (2015) used the information quantity method and logistic regression to assess the geodisasters susceptibility in small-scale geomorphological units. Fu (2008) constructed a SVM model and used GIS technology to assess the landslide susceptibility based on indices of slope, rainfall and vegetation coverage as the model inputs. Feng et al. (2016) developed an ANN model based on GIS raster analysis to select the related environmental factors including slope, curvature and vegetation, thereby generated the landslide susceptibility zoning map. Pham et al. (2016)

proposed an ensemble method integrating a hybrid rotation forest and fuzzy disordered rule induction classifier for landslide detection. Ermini et al. (2005) also used the ensemble method by developing an ANN based on multilayer perceptron (MLP) and probabilistic neural network (PNN). Oh et al. (2018) used SVM to assess the landslide susceptibility with 18 affect factors, and the relationships between landslide occurrence and these factors were analyzed.

Although having better performance, machine learning methods can only obtain limited representative features from the data, have poor generalization ability, and may easily cause local optimum, all of which limit the further development of the machine learning methods in this field.

**Geodisasters Susceptibility Zoning with Deep Learning**

Deep learning has gained great achievements in many fields, such as image segmentation, target detection, and natural language processing. In recent years, the quality and quantity of geodisasters data have increased dramatically, with characteristics of multi-source, spatio-temporary and non-linear, etc., which requires higher ability for massive data analysis. Research has gradually demonstrated the potential of deep learning in effectively analyzing geodisaster data. It has been introduced for many practical tasks, such as detections of landslide and debris flow, landslide susceptibility assessment, seismic data interpolation, denoising, and earthquake detection and localization.

Huang et al. (2022) proposed a multi-channel weighted, layer adaptive classification LAW-CNN architecture based on the information entropy of the landslide impact factor, and applied it for landslide susceptibility mapping. Jing et al. (2022) used depth belief network (DBN) for landslide susceptibility assessment with topographic and geomorphologic features, geological features, hydrological features, and human activities as the model inputs. Comparison of the DBN with SVM, BP neural network, and logistic regression indicated a much higher performance of DBN for this task. Wang et al. (2019) used multicollinearity analysis and gain ratio (GR) method to extract and select features for the CNN-based landslide susceptibility assessment. Wang et al. (2020a, 2020b) proposed a DBN model containing three layers of restricted Boltzmann machine (RBM) for the susceptibility assessment. Natural breakpoint method was applied in the classification of susceptibility indices. Wang et al., (2020a, 2020b) used multicollinearity analysis to extract and select features, which were thereby weighted and ranked. Features with the higher weight were input into RNN in priority, so that more critical information could be retained. Besides, Wang et al. (2021) proposed a LSTM-CNN framework for landslide susceptibility evaluation. Yi et al. (2020) proposed a CNN model based on multiscale sampling to select spatial information around landslide and non-landslide points to generate datasets in different scales. Li et al. (2022) constructed an ensemble method integrating CNN and RNN for the landslide susceptibility assessment.

### 6.1.3 Research Object and Contents

The mechanism of geodisasters is complex and many factors can influence the susceptibility. Deep learning models with their deeper and more complex architectures can analyze the occurrence pattern of geodisasters in depth and often achieve more accurate results in geodisaster susceptibility analysis. In addition to the CNN-based models and RNN-based models, graph convolutional network (GCN) attracts more attentions in recent years. Most of the current DL-based susceptibility assessing methods only treat the geodisasters data as Euclidean structural data, ignoring the potential inherent topological relationships. The GCN can learn structural features and extract the structural relationships between neurons, which helps to improve the accuracy of susceptibility analysis results.

Thus, this chapter constructs a GCN-based model to assess the landslide susceptibility in the study area of Yanbian Korean Autonomous Prefecture, China. The details of the study are as follows:

(1) Basic geological and topographic data including the geological structure, topography and geomorphology, climate and hydrology of the study area are collected. Pearson correlation coefficient is used to select more related features, including the elevation, slope angle, slope aspect, land use, distance to roads, distance to water, stratigraphic lithology and fault.
(2) Raster pixel is regarded as the unit of assessment. ArcGIS is used to spatially combine the feature layers. Features of landslide area and non-landslide area are extracted from these layers and converted to 2D form to construct the dataset.
(3) This chapter constructs a GCN-based geodisaster susceptibility assessing model, extracts high-dimensional features using CNN, introduces depth separable convolution to extract spatial features and channel features, and performs graph convolution on them, respectively.
(4) The model is trained and used to predict the pixel-based landslide susceptibility index. Natural breakpoint method is applied in the zoning, and the landslide susceptibility map is generated for the study area.

The technology roadmap of this research is shown in Fig. 6.1.

The arrangement of each section is: Sect. 6.2 introduces the principle of methods and techniques used in this study, including GCN and CNN. Section 6.3 describes the geographical location, meteorology and hydrology, topography and geomorphology, geological conditions and human engineering activities of the study area. Section 6.4 is about the whole process for the construction of the geodisaster susceptibility assessing model based on GCN, evaluation of model performance and the prediction of the susceptibility in the study site. Section 6.5 summarizes the research and proposes improvement suggestions for the problems of current research method.

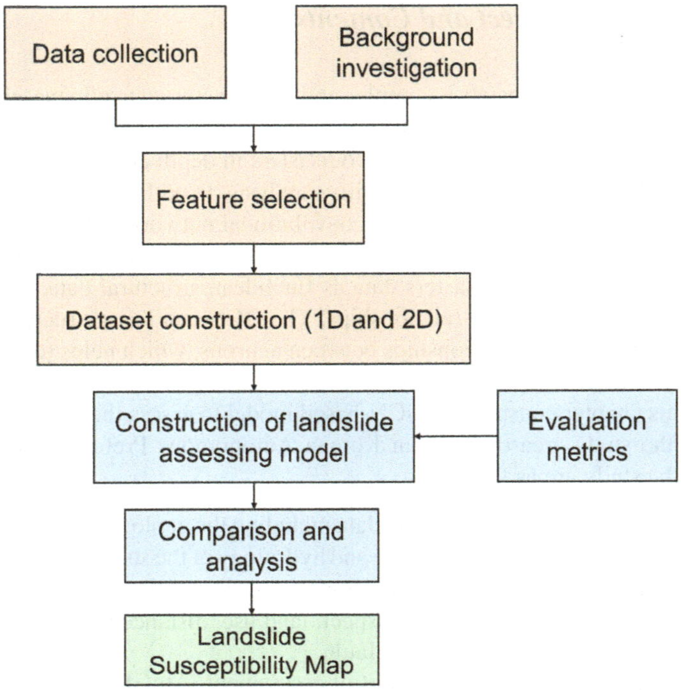


**Fig. 6.1** Flow chart of the study

## 6.2 Related Principles and Techniques

### *6.2.1 Overview of the Graph*

Deep learning stands out in many fields for its ability to extract hidden features from Euclidean structured data. For example, CNN can learn to share convolutional kernels to build local connections and locally extract features from images represented in Euclidean structure. However, many data in reality cannot be represented by Euclidean structure, but only be represented by graphs, such as social networks and knowledge graphs. This leads to the convolution operation being hard to implement in irregular graph structures. In this case, graphs begin to be used to represent non-Euclidean structured data. Graph is a non-Euclidean data structure consisting of edges and nodes that can be used to represent relationships and store data. All of the edges and nodes of a graph have their own explicit meanings. According to whether the edges are directed or not, graphs can be divided into directed graph and undirected graph. The relationship between nodes is expressed based on edges, and the magnitude of the relationship can be further expressed in terms of weights. In this study, an undirected graph is used, and its structure is shown in Fig. 6.2.

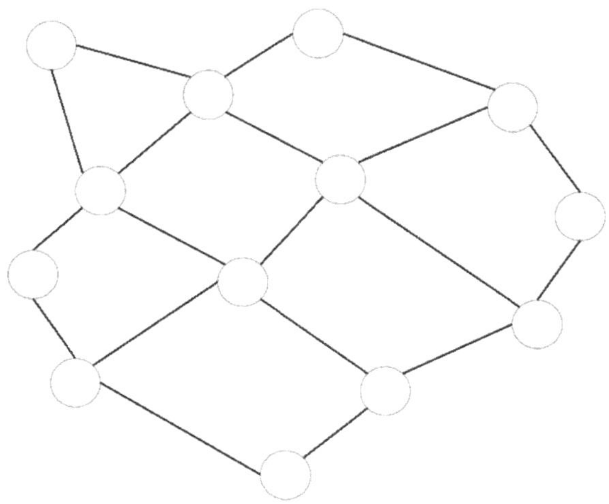

**Fig. 6.2** Undirected graph

The nodes of a graph are represented by $V_i \in V$. $E$ indicates the set of connections between the nodes of the graph, and $G = (V, E)$ represents a graph. The edges in the graph structure are represented by $E_{i,j} \in E$, which is the connection between nodes $i$ and $j$. In computer science, adjacency matrix is usually used to store the defined graph and then process the graph data. The adjacency matrix uses a 2D matrix to represent the connection relationship of edges, which means the construction of matrix $|V| \times |V|$ can gain the adjacency matrix of the graph. If the nodes $V_i$ and $V_j$ are connected, $A_{ij} = 1$.

### *6.2.2 Graph Convolutional Network*

Graph Neural Network (GNN) (Gori et al., 2005) is a DL-based method for learning graph data that firstly proposed by Gori et al. (2005). GNN works based on graph structures and can learn entities, relational vector representations, and composition rules in graph data, and is widely used because of its excellent information aggregation ability. In recent years, researchers have focused on using the deep learning models to process graph data. Bruna et al. (2013) introduced convolution in GNN for the first time and proposed a frequency domain-based GCN, but it is of high time complexity. To tackle this issue, Kipf and Welling (2016) proposed a GCN based on spatial domain, which allowed the model to be applied in large-scale graph data processing tasks. In frequency domain-based GCN, the graph convolution operations in the network are performed using the Laplace transform in the frequency spectrum. In the spatial domain-based GCN, graph convolution is carried out based on the relationships between nodes in space.

Inspired by 2D convolution, graph convolution is proposed. Figure 6.3 illustrates the operation of 2D convolution and graph convolution. In the 2D convolution, each

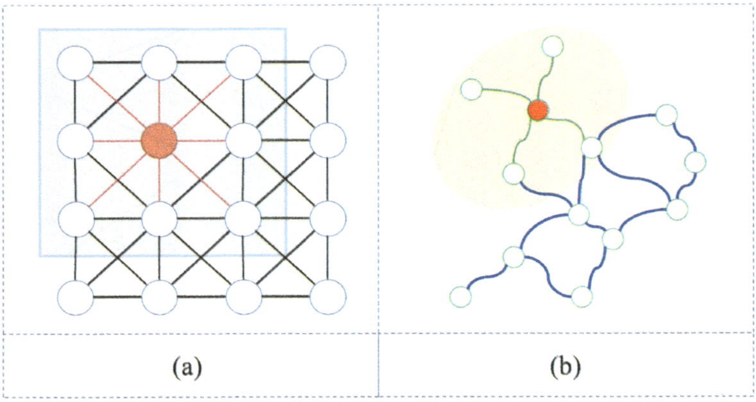


**Fig. 6.3** 2D convolution (**a**) and graph convolution (**b**), modified from Wu et al., (2020a, 2020b)

pixel in an image is considered as a node. The neighboring nodes are determined by the size of the convolution kernel and the values of neighboring nodes are fixed for any single node. The 2D convolution shown in Fig. 6.3a uses the weighted average value of the red node and its neighboring nodes. Unlike the 2D convolution, the neighboring nodes of a node in the graph convolution are unordered and of various values. The graph convolution shown in the Fig. 6.3b takes the average of the eigenvalues of the red node and its neighboring nodes.

GCN can extract features from graph structured data by constructing adjacency matrix, which promotes the definition of neighborhoods and can combine multi-scale information to form high-level representations. GCN can effectively utilize graph structure information and attribute information (Wu et al., 2020a, 2020b). Equation (6.1) represents the graph convolution (Wu et al., 2020a, 2020b):

$$H^{(l+1)} = f\left(H^{(l)}, A\right) = \sigma\left(\tilde{D}^{-\frac{1}{2}}\tilde{A}\tilde{D}^{-\frac{1}{2}}H^{(l)}W^{(l)}\right) \tag{6.1}$$

where $\tilde{A} = A + I$, $I$ is the identity matrix. $\tilde{D}$ is the degree matrix of $\tilde{A}$. $W^{(l)}$ is the weight matrix. $\sigma$ denotes a nonlinear activation function. $X$ is the input, so the input layer can be expressed as $H^{(0)} = X$. $Z$ is the output, so the output layer can be expressed as $H^{(L)} = Z$, and $L$ is the number of layers. Different GCN will have different $f()$ functions.

Figure 6.4 shows the process of implementing GCN for the image classification. The input is the graph data and its adjacency matrix. The data is processed by the graph convolution operation in each hidden layer and delivered to the following layer. In the fully connected layer, the graph data is transformed into feature vectors and the final classification result is output by using the softmax function.

Like CNN, GCN can share local parameters and the receptive field can enlarge with the increasing number of layers. The more layers, the larger the receptive field and the more information. What different from CNN is that GCN can be used to

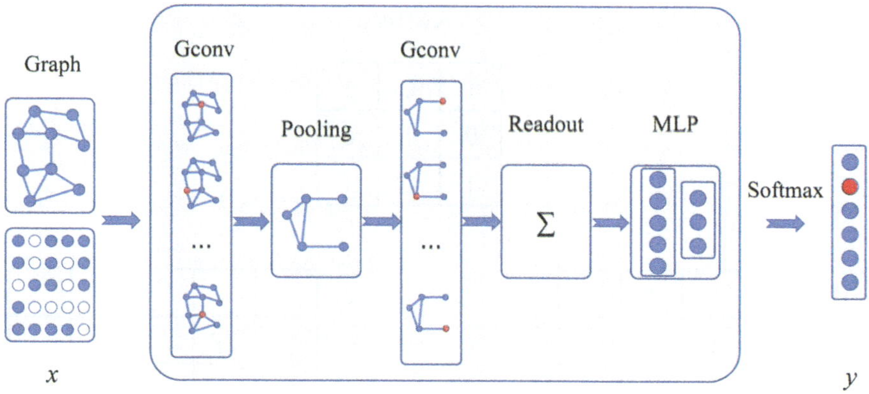


**Fig. 6.4** GCN classification process, modified from Wu et al., (2020a, 2020b)

process unstructured data. In addition, the edges of the graph data represent the dependency relationship between nodes in GCN, which is only a property of nodes in CNN. And GCN can use this dependency relationship to aggregate node features.

### 6.2.3 *Convolutional Neural Network*

LeNet-5 was firstly proposed by Lecun et al. (1998). It was the first CNN to be applied in digital recognition. The main structures of LeNet-5 include convolutional layer, activation function, pooling layer and fully connected layer.

**Convolutional Layer**

The convolutional layer plays a key role in the CNN. Receptive field and shared weights are the two important parts of convolutional layer. The receptive field describes the size of the region where the pixels of the original image are mapped in the feature map of the CNN. As the regional features are more closely connected, the features extraction of CNN starts from regional scale. Neurons in the hidden layer use the receptive field with fixed size to extract features in regional scale, and then higher-level structures and these regional features are combined to get the global features of the input data. The receptive field can be expressed in terms of central location and size. And the central part of the receptive field usually contributes more to the output features. The size of the receptive field can be customized, and usually there is only one convolutional kernel for a receptive field. The value in the convolutional kernel continuously changes during training, which is also a parameter in the CNN. When the convolution kernel moves on the input data with a specific stride size, the corresponding feature map is computed by convolution, and the local features are extracted. The convolution kernels are used with same size and weights in the same

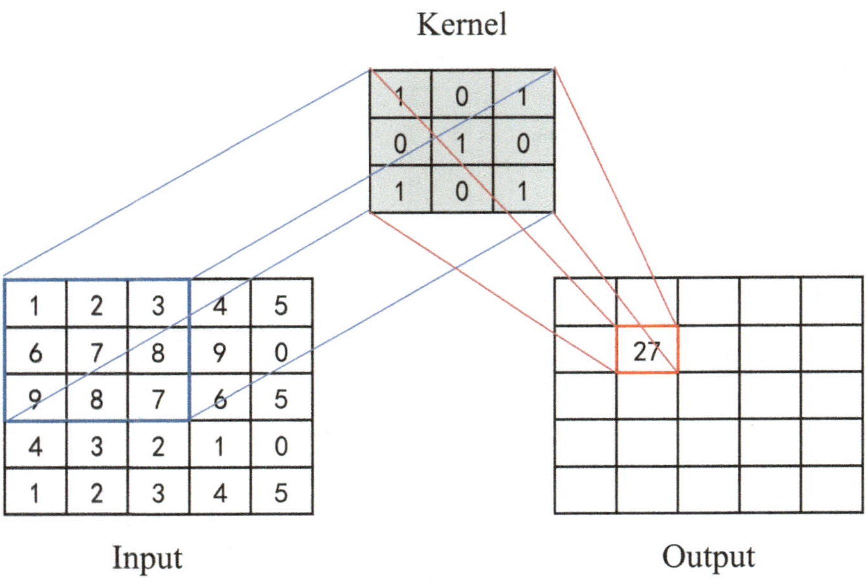


**Fig. 6.5** Convolution process

feature map, which are called "shared weights". The convolution process is shown in Fig. 6.5.

A convolutional layer contains several convolutional kernels. The convolution process is performed with the feature map obtained from the previous layer and the convolution kernel. Then the result of convolution process is calculated by the activation function to obtain the final output feature map. The feature map formed by the convolution kernel $k$ of the layer $i$ is defined as $Z_i^k$, the weight matrix is defined as $W_i^k$, bias is $b_i^k$, activation function is $f$, so the convolution operation can then be expressed in Eq. (6.2):

$$Z_i^k = f\left(W_i^k \otimes Z_{i-1}^k + b_i^k\right) \tag{6.2}$$

**Pooling Layer**

CNN uses convolutional layers to extract features from the data, but this can lead to problems such as information redundancy and overfitting. Pooling layers between successive convolutional layers can reduce the dimensionality of the data, compress the input feature matrix, and also filter out some of the non-dominant feature information. These benefits allow the data to be represented by higher-level features, improve the computational efficiency and reduce overfitting to some extent, which are helpful to optimize the model.

The pooling layer uses a pooling filter to scan the feature map. The most common pooling operation is maximum pooling, which takes the maximum eigenvalue of

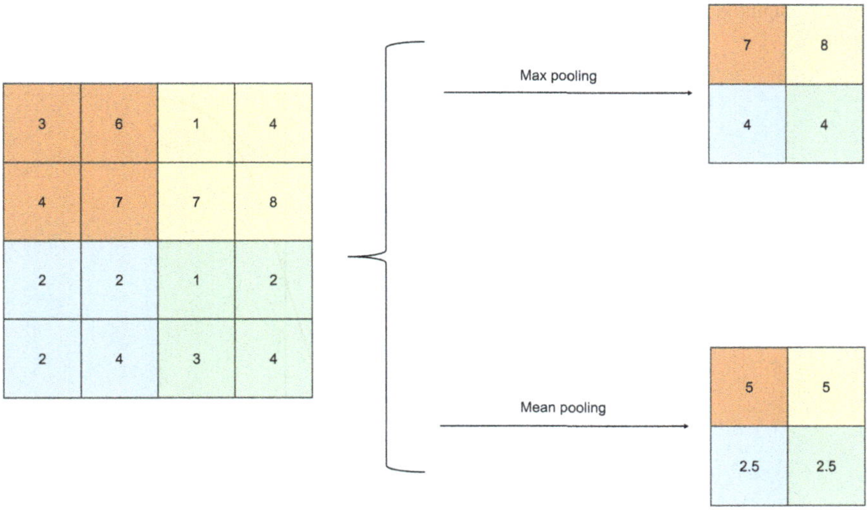


**Fig. 6.6** Maximum pooling and average pooling operations

the feature map within the size of a pooling filter as the output. Average pooling is also commonly used which takes the average eigenvalue of the feature map as the output. Maximum pooling and average pooling operations are shown in Fig. 6.6. The pooling operation focuses on if the feature existed without caring its location, which allows the main features to be preserved.

### Activation Function

The convolutional and pooling layers of a CNN can solve linear problems, while the nonlinear factors introduced by the activation function allows the model to solve nonlinear problems. The common activation functions include:

(1) Sigmoid function.

Sigmoid functional equation can be expressed in Eq. (6.3) and the image is shown in Fig. 6.7.

$$s(x) = \frac{1}{1 + e^{-x}} \tag{6.3}$$

The sigmoid function can map real numbers between 0 and 1. It normalizes the output and is suitable for models for outputting predicted probabilities. However, for very large or very small input values, the sigmoid function tends to saturate and the gradient converges to zero. In backward propagation, the derivative also converges to 0 for neurons whose output are close to 0 or 1, resulting in that the weights cannot be updated and features cannot be extracted. In addition, the output of the sigmoid function is not zero-centered. When the inputs of the post-layer neurons are

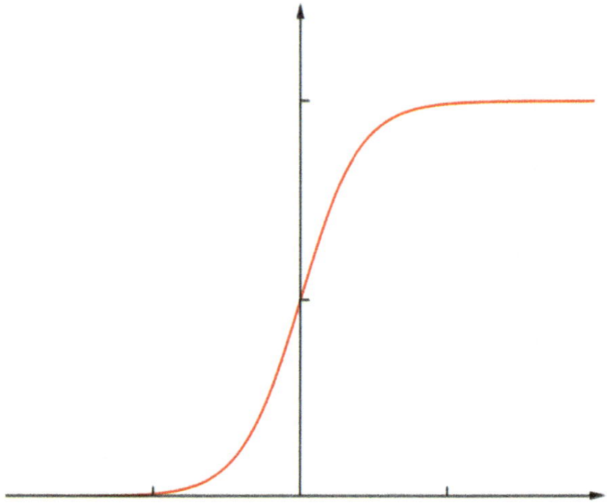

**Fig. 6.7** Sigmoid activation function

all positive, the local gradients of the weights are also all positive, leading to a slower convergence in model training.

(2) Tanh function.

Tanh equation can be expressed in Eq. (6.4) and the image is shown in Fig. 6.8.

$$T(x) = \frac{1 - e^{-2x}}{1 + e^{-2x}} \tag{6.4}$$

The tanh function is also known as the hyperbolic tangent activation function. The output of the tanh function is between − 1 and 1. The derivative of the tanh function

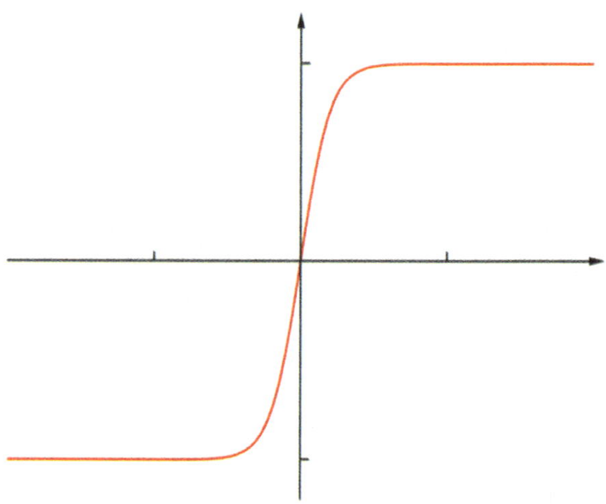

**Fig. 6.8** Tanh activation function

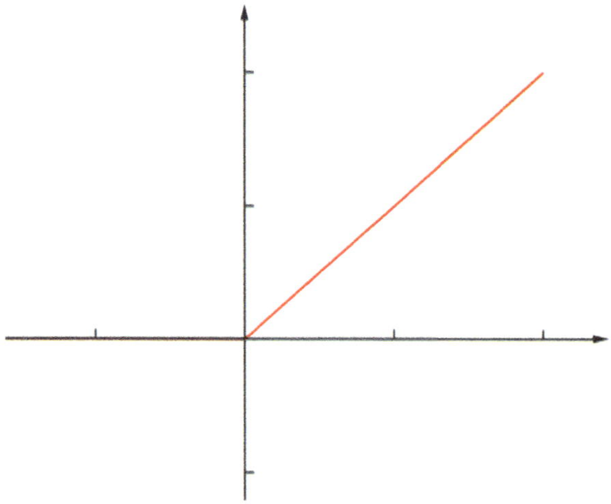

**Fig. 6.9** ReLU activation function

tends to zero when the input value is very large and small. Tanh activation function still suffers from the problem of gradient saturation.

(3) ReLU function.

ReLU functional equation can be expressed in Eq. (6.5) and the image is shown in Fig. 6.9.

$$R(x) = \max(0, x) \tag{6.5}$$

The ReLU function is a segmented linear function that has been widely used in deep neural networks. If the input is a value lower than 0, the output value is equal to 0. If the input is a value > 0, the output value grows linearly. This property tackles the issue of gradient disappearance in sigmoid and tanh functions. When the input value is > 0, the derivative is 1, which accelerates the convergence of the gradient descent and requires only a threshold to be activated. When the input is lower than 0, the output value is equal to 0. In this case, only parts of neurons are activated, giving the model a better ability of generalization.

The ReLU function is not zero-centered, which can lead to slower convergence during the model training. And if the parameters are not updated incorrectly during the training, the gradient of a certain neuron will be zero and will not active the training data.

(4) Leaky ReLU function.

Leaky ReLU functional equation can be expressed in Eq. (6.6) and the image is shown in Fig. 6.10.

$$LR(x) = \begin{cases} ax, & (x\langle 0, a\rangle 0) \\ x, & (x \geq 0) \end{cases} \tag{6.6}$$

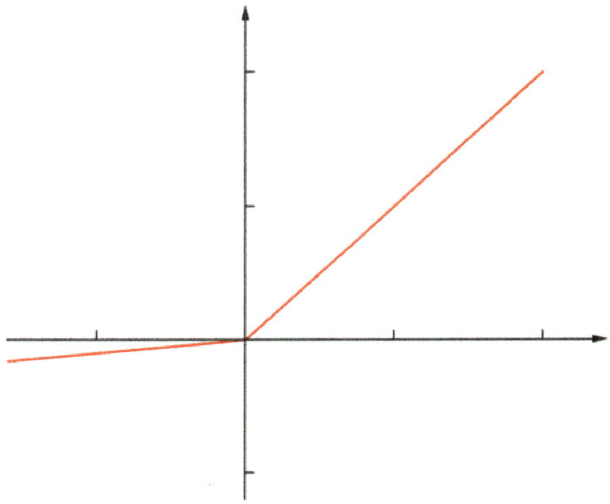

**Fig. 6.10** Leaky ReLU activation function

The Leaky ReLU function solves the problem that the ReLU function is ineffective when the input value is lower than 0. When this happens, the output is a linear function with a small slope, which can preserve parts of the negative values and active parts of the neurons of the ReLU function.

**Fully Connected Layer and Softmax Layer**

The fully-connected layer acts like a classifier in the model. While the convolutional layer extracts local features, the fully connected layer uses a matrix of weights to recombine the local features and transform the feature map to an 1D vector. It enhances robustness by ignoring the properties of the spatial structure and voting on all the previously extracted features.

The Softmax layer, usually at the last layer in the network, is used to calculate the probability of each category, and the determined classification is the category with the highest probability. The softmax function is used in multi-classification problems, which is expressed as Eq. (6.7):

$$softmax(z_i) = \frac{e^{z_i}}{\sum_{j=1}^{n} e^{z_j}} \tag{6.7}$$

where $z$ is the output value of the fully connected layer and $n$ is the number of categories.

## 6.3 Study Site

### *6.3.1 Environmental Conditions*

The study site is in the Yanbian Korean Autonomous Prefecture, Jilin Province, China (129°40′48″ ~ 130°00′00″ E, 43°00′00″ ~ 43°13′12″ N). The total area of Yanbian Korean Autonomous Prefecture is 43,300 $km^2$.

**Meteorology and Hydrology**

Yanbian Korean Autonomous Prefecture is located in the middle temperate zone of the Northern Hemisphere and has a humid monsoon climate. Since it belongs to the Changbai Mountains which has a complex topography and is adjacent to the Japan Sea, the climate here has obvious vertical distribution characteristics. The lower reaches of the Tumen River and the middle and lower reaches of the Hunchun River belong to the maritime climate zone, with strong winds in spring, rainy days in summer and high temperatures in winter. The basins of the Hailan River and the Burkhartung River basin have sufficient heat and moisture. The Dunhua Plateau, located in the western part of the Halba Ridge, has a humid and cold climate. The rest of the region belongs to the Changbai alpine region with an average annual sunshine of 2150–2480 h and an average temperature of 2–6 °C. As altitude and latitude rise, the average temperature tends to decrease.

Yanbian Korean Autonomous Prefecture is rich in water resources. There are 487 rivers with a total length of 10,900 km, including 470 rivers with the area higher than 20 $km^2$ and 137 rivers with area higher than 100 $km^2$. The total average annual flux is about 7.31 billion cubic meters. There are also 9 natural lakes with area higher than 1 $km^2$.

**Topography**

The study area is in the low mountainous hilly area, and the terrain is higher in the northwest and lower in the southeast (Fig. 6.11). The highest elevation is 977.4 m in Mount Mopan and the lowest point with an elevation of 16 m is a river valley plain area formed with Gaya River. The average elevation of the study area is 360.15 m, with a relative height difference of 960 m. The landforms of the study area include valley plain area, low mountainous area and hilly area. The low mountainous area is in the northeast of the region with abundant shrubs. The hilly area is in southwest of the region, mainly for agricultural activities. The valley plain area is along the river banks of Gaya and Bratong Rivers, which is a concentration of towns and villages.

**Stratigraphic Lithology**

The stratigraphy of the study area is mainly consisted by sedimentary and intrusive rocks. Due to the long-term denudation, the distribution area of ancient strata is small, while the Mesozoic strata are more developed. The exposed rocks are complex, mainly from Carboniferous, Permian, Triassic, Jurassic, Cretaceous, Tertiary and

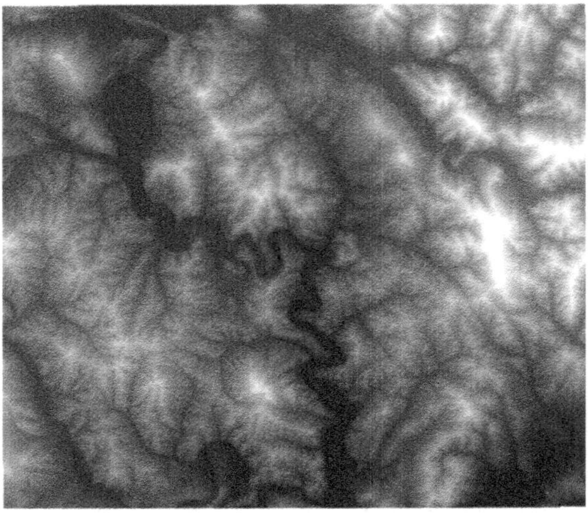

**Fig. 6.11** DEM map of the study area

Quaternary. The strata lost can be found in the study area and there are unconformable contacts in the strata.

**Neotectonics**

The study area is in the eastern part of the Central Asian orogenic belt, connecting the southern edge of the Xingmeng orogenic belt and the northern edge of the North China Platform. It belongs to the eastern Yanbian Fold Belt of the Jihei Fold System. Influenced by the tectonic evolution from early Paleozoic to Cenozoic, the study area formed NE-trending and NS-trending tectonic tracks and a basin-ridge tectonic system.

The strong folding and uplifting in most parts of the study area caused regional metamorphism and magma intrusion activities. The intrusion structures form the current profiles of Nangang Mountains, the Ying'er Ridge and the Uliangha Mountains. After the Mesozoic, the tectonic activity in the region was mainly the differential uplift of fault blocks, forming the Mesozoic clastic basins in Cretaceous and Jurassic strata. By the Cenozoic, very few isolated basins deposited Tertiary clastic rocks. Due to the neotectonics, long-term weathering and denudation, the Cretaceous clastic basins in the study area were slowly uplifted, forming the present landscape of gullies, valleys and rolling hills.

## 6.3.2 *Human Activities*

With the increasing demand of economic development, water conservancy and hydropower projects in the study area are expanded. Mining and infrastructure construction are becoming more frequent, leading to an increasing possibility of geodisasters. According to the report of the Department of Natural Resources of Jilin Province (2019), there were 2860 geodisaster potential sites (including 120 important potential sites) in Yanbian Korean Autonomous Prefecture in 2019, threatening a total of 12,087 people and total assets of about 782 million RMB.

## 6.4 Model Construction

### *6.4.1 Depthwise Separable Convolution*

Standard convolution can learn both spatial features and channel features of the input feature map by convolution operations. Unlike standard convolution, depthwise separable convolution (Howard, et al., 2017) separate a complete convolution operation into depthwise convolution and pointwise convolution, which decouples the spatial correlation of the convolution layer from the channel correlation. This reduces the number of parameters and keeps the accuracy loss within an acceptable range.

Figure 6.12 indicates the depthwise convolution process and pointwise convolution process. It is supposed that number of channels is N for the input data. First, the depthwise convolution operation is performed on the input data, which is separated into multiple single-channel feature maps. The convolution operation is performed on each single-channel feature map, and then the results are stacked. The size of the convolution kernel can be defined as. The feature map generated by the depthwise convolution will have the same number of channels as that of the input data. Then, the pointwise convolution is performed on the feature map. The filter contains same amount of $1 \times 1$ convolution kernels with the number of channels. In this process, the convolution operation calculates the feature map in a depth-weighted manner to form a new feature map. The number of channels of the output feature map is determined by the number of filters.

### *6.4.2 Model Structure*

In this research, a DL-based geodisaster susceptibility assessing model is constructed, which consists of CNN module, GCN module, fully connected layer and softmax layer. Firstly, the raw data is transformed to the 2D input data for the feature extraction. CNN module is used to extract high-dimensional features. These high-dimensional features are then processed by using the depthwise separable convolution to extract spatial features and channel features. The adjacency matrices are constructed for graphs convolution, respectively. Finally, the features are integrated by the fully connected layer, and a softmax layer is used to output the classification results of landslide and non-landslide pixels with their probability values, which is considered as the landslide susceptibility index.

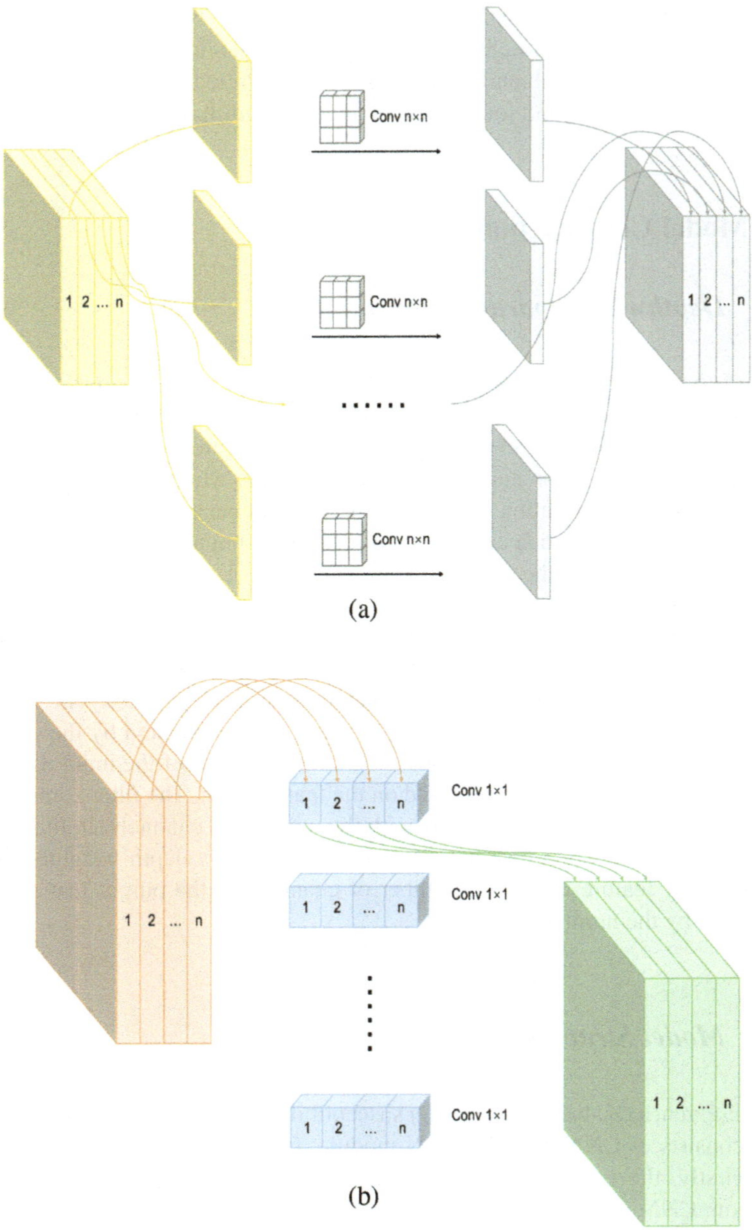


**Fig. 6.12** Depthwise convolution process (**a**) and pointwise convolution process (**b**)

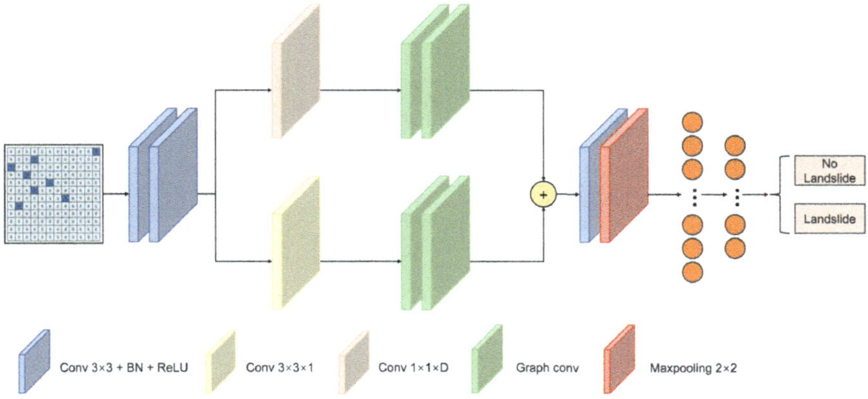


**Fig. 6.13** DL-based model structure for geodisaster susceptibility assessment

The model structure is shown in Fig. 6.13. The input data is defined as $X \in R^{N \times D}$, $D$ is the number of the channels and $N = H \times W$ is the size of the 2D matrix.

In this study, the CNN module contains two convolutional layers. Each convolutional layer is padding with 0, and has the size of 3 × 3 and the stride of 1. The ReLU function is used as the activation function. After the convolution operation, the features $X_1 \in R^{H \times W \times D_1}$ are obtained, where $D_1$ is the number of channels changed after the convolution operation.

Depthwise separable convolution is performed in the GCN module. In depthwise separable convolution, convolution is performed separately for each channel of the feature map to obtain spatial features $X_s$ by using the k × k × 1 filter. Similarly, to obtain the channel features $X_c$, the features are convolved by using the $1 \times 1 \times D_1$ filter. The graph structure is constructed based on spatial features $X_s$ and channel features $X_c$. In both $X_s$ and $X_c$, the neurons whose value change significantly contain more information and make it more reasonable to construct edges between these nodes. In this basis, spatial adjacency matrix $A_s \in R^{D_1 \times D_1}$ and channel adjacency matrix $A_c \in R^{(N \times N)}$ can be constructed. As all the selected factors have been proved can contribute to the landslide occurrence, the inherent characteristics of these factors are assumed to be interconnected, so the constructed graph is fully connected. Therefore, the parameters of $A_s$ and $A_c$ are considered as 1 here. Afterwards, the input spatial features and channel features are graphically convolved separately, and the results are summed to for feature fusion. From Eqs. (6.1), (6.8) and (6.9) are obtained. Figure 6.14 shows the process of GCN module feature extraction.

$$H_s^{(l+1)} = \sigma\left(\tilde{D}_s^{-\frac{1}{2}} \tilde{A}_s \tilde{D}_s^{-\frac{1}{2}} X_s^{(l)} W_s^{(l)}\right) \tag{6.8}$$

$$H_c^{(l+1)} = \sigma\left(\tilde{D}_c^{-\frac{1}{2}} \tilde{A}_c \tilde{D}_c^{-\frac{1}{2}} X_c^{(l)} W_c^{(l)}\right) \tag{6.9}$$

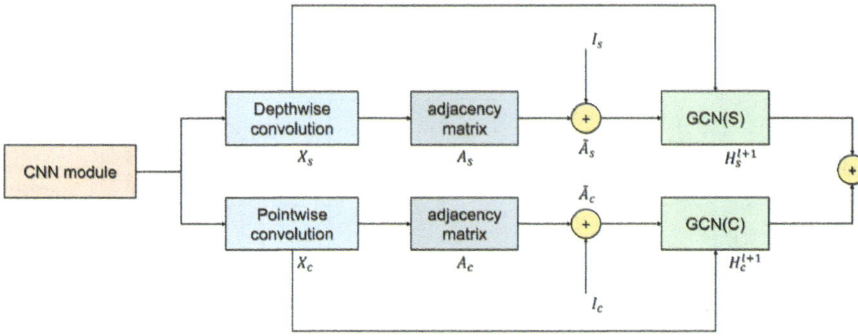


**Fig. 6.14** GCN model

The features obtained from the GCN module are convolved and downsampled, where the pooling kernel size of the maximum pooling layer is $2 \times 2$ and the stride is 1. The features calculated in the previous module are mapped into the label space by the fully connected layer, meanwhile the fully connected layer converts the features extracted by the GCN module into 1D vectors by combining the weight matrix. The softmax layer finally predicts the probabilities of classification categories.

### *6.4.3 Feature Selection*

The selection of geodisaster related features is a key step in the study of geodisaster susceptibility (Reichenbach et al., 2018). In this research, eight features related to the landslide susceptibility were selected, including elevation, distance to road, distance to water, slope angle, slope aspect, land use, fault and stratigraphic lithology. Feature maps are generated by using ArcGIS, and all the maps are resampled to 5 m resolution to obtain the raster data for each feature.

Strong correlation between features can cause data redundancy and affect the model performance. Pearson correlation coefficient method is used test the correlation between the features. The absolute value of correlation coefficient in the range of 0.7 and 1 indicates a strong correlation, in the range of 0.3 and 0.7 indicates a moderate correlation, and in the range of 0 and 0.3 indicates a weak correlation. Equation (6.10) is for the Pearson correlation coefficient calculation (Zhang et al., 2005):

$$Pearson = \frac{\sum_{i=1}^{n}(x_i - \overline{x})(y_i - \overline{y})}{\sqrt{\sum_{i=1}^{n}(x_i - \overline{x})^2}\sqrt{\sum_{i=1}^{n}(y_i - \overline{y})^2}} \tag{6.10}$$

where x and y are variables and $\overline{x}$ and $\overline{y}$ are variable means.

**Table 6.1** Pearson correlation coefficients

| Features | Lithology | Distance to road | Land use | Distance to water | Slope aspect | Slope angle | Elevation | Fault |
|---|---|---|---|---|---|---|---|---|
| Lithology | 1 | – 0.08 | 0 | – 0.40 | – 0.16 | – 0.11 | 0 | 0.09 |
| Distance to road | | 1 | 0 | 0.28 | 0.41 | 0.22 | 0 | 0.07 |
| Land use | | | 1 | 0.13 | 0 | – 0.16 | – 0.07 | 0 |
| Distance to water | | | | 1 | – 0.09 | 0.15 | 0.20 | 0.40 |
| Slope aspect | | | | | 1 | 0.14 | 0 | 0.13 |
| Slope angle | | | | | | 1 | 0 | 0.13 |
| Elevation | | | | | | | 1 | 0 |
| Fault | | | | | | | | 1 |

The Pearson correlation coefficients of selected features are shown in Table 6.1. The absolute values of the correlation coefficients are all $< 0.5$.

## 6.4.4 *Model Training*

### Experimental Environment

In this experiment, the Python was chosen for programming and the Pytorch framework was used to construct the model. The hardware environment used is shown in Table 6.2 and software environment used is shown in Table 6.3.

### Experimental Data

The raster pixel can be regarded as a 1D vector containing the feature. In this basis, the vector is converted into 2D data of $n \times n$ ($n$ is the maximum value of all features). Therefore, the 1D data with 8 features are expanded into a matrix of $12 \times 12$, as shown in Fig. 6.15.

**Table 6.2** Hardware environment

| Hardware | Parameter |
|---|---|
| CPU | Intel(R) Core (TM) i3-10100F@ 3.60GHz |
| GPU | NVIDIA GeForce GTX 960 |
| Video memory | 4G |
| Memory | 16G |

**Table 6.3** Software environment

| Software | Parameter |
|---|---|
| Operation system | Windows 10 64-bit |
| Cuda | Version 10.2 |
| Platform | JetBrains PyCharm 2019 |
| Deep learning framework | Pytorch |

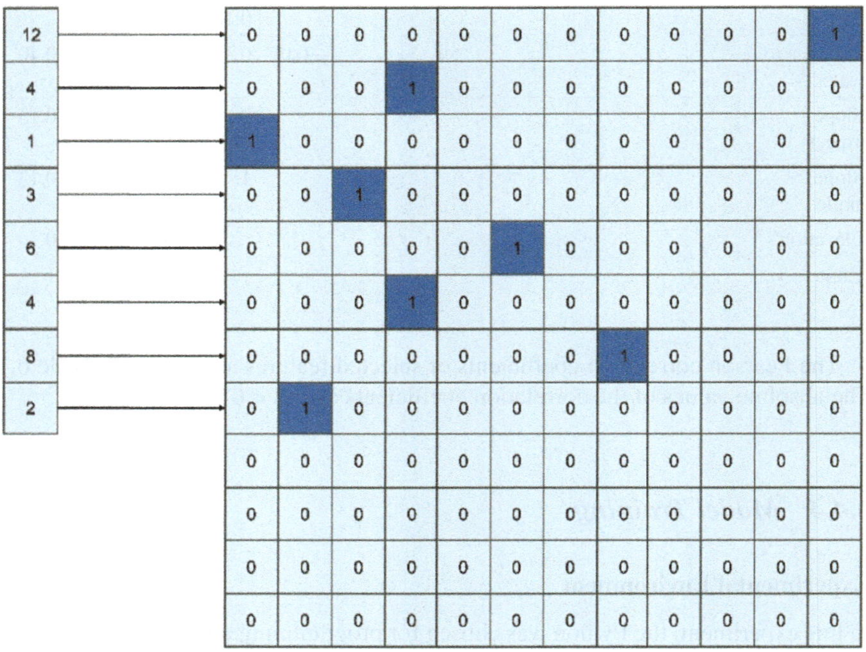


**Fig. 6.15** Data conversion

The prepared dataset is divided into train set, validation set and test set, which contain both positive and negative samples. There are 33,218 landslide pixels in the study site to be the positive samples, and another 33,218 pixels are randomly selected from the non-landslide area to be the negative samples. Overall, 60% data is for the training, 20% data is for the validation, and 20% data is for the test.

### Hyperparameter Setting

The trial-and-error method is used to fine-tune the hyperparameters and obtain the relatively optimal results for landslide susceptibility analysis. During the training process, neurons are discarded with a probability of 0.5 in the Dropout layer and the batch size is set to 10. When the initial learning rate is set to 0.001, the loss can decrease faster with the higher accuracy. The learning rate is updated with a momentum of 0.9 after every 5 iterations of training, and the maximum number of iterations is set to 500. The learning rate of the gradient descent method is optimized using Adam algorithm. The loss function uses the cross-entropy function, which is calculated as shown in Eq. (6.11):

$$Loss = -\sum_{i=1}^{N} y_i \log p_i \tag{6.11}$$

where N denotes the total number of categories, and in this experiment N = 2 (landslide and non-landslide). $y_i$ is the indicator variable, $y_i = 0$ or 1. When the category of the sample is the same as the predicted category, $y_i = 1$, otherwise $y_i = 0$. $p_i$ denotes the probability of a sample being predicted into the $i$ category.

### 6.4.5 Results

#### Evaluation Metrics

In this study, accuracy, precision, recall, f1-score and receiver operating characteristic curve (ROC curve) are used to evaluate the model performance.

Accuracy refers to the ratio of the number of samples correctly predicted by the model to the total number of samples. Accuracy can be calculated by Eq. (6.12).

$$Accuracy = \frac{TP + TN}{TP + TN + FP + FN} \tag{6.12}$$

Precision refers to the ratio of the number of samples predicted by the correct model to the number of all samples predicted to be positive. Precision can be calculated by Eq. (6.13).

$$Precision = \frac{TP}{TP + FP} \tag{6.13}$$

Recall refers to the proportion of all positive samples that are predicted correctly. Recall can be calculated by Eq. (6.14).

$$Recall = \frac{TP}{TP + FN} \tag{6.14}$$

Considering the precision and recall are in a mutually constraining relationship, f1-score is used to weigh precision and recall. F1-score can be calculated by Eq. (6.15).

$$F1Score = \frac{2TP}{2TP + FP + FN} \tag{6.15}$$

where true positive (TP) indicates the number of positive samples correctly predicted to be the landslide area. False positive (FP) is the number of negative samples wrongly predicted to be the landslide area. True negative (TN) indicates the number of negative samples correctly predicted as non-landslide area; False negative (FN) is the number of positive samples wrongly predicted as non-landslide area.

The receiver operating characteristic curve (ROC curve) is also used to evaluate the model performance. It calculates a series of susceptibility and specificity by defining thresholds and plots the curve. The model performance can be indicated by the area under the curve, which is the AUC value. The higher AUC value means a better performance of the model.

**Model Performance**

CNN-1D model (Wang et al., 2019) and DBN model (Wang et al., 2020a, 2020b) are used to evaluate and compare with proposed GCN model performance. The CNN-1D model expresses the raw data extracted from the feature map in a one-dimensional form and uses CNN as the architecture to assess the landslide susceptibility. Figure 6.16a illustrates the CNN-1D model structure. DBN model consists of three restricted Boltzmann machines, which are firstly pre-trained before the hyper-parameter optimal training. Then the trained model is used to make predictions and obtain landslide susceptibility assessing results. Figure 6.16b illustrates the DBN model structure.

All the models are trained and tested in same dataset with same settings. Above-mentioned evaluation metrics are used to compare the performance of GCN model with DBN model and CNN-1D model. The results are shown in Table 6.4 and the highest values of evaluation metrics are bolded.

The results indicated that the GCN model has the highest overall accuracy. Comparing with CNN-1D model, the precision of proposed model is improved significantly by 4.83%, and accuracy and f1-score are also slightly improved by 1.57 and 1.27%, respectively. Similarly, comparing with DBN model, the precision of the proposed model is improved by 5.26%, and accuracy and f1-score are increased by 2.27% and 1.99%, respectively. Figure 6.17 shows the ROC curves of the three models. Comparing with DBN and CNN-1D models, the AUC values of GCN are higher and have better prediction ability for this dataset.

**Landslide Susceptibility Analysis**

The trained models are used to predict on pixels to obtain the probability that pixel is a landslide area. The probability value is between 0 and 1 and is also called the susceptibility index. The susceptibility is classified into five classes of very low susceptibility, low susceptibility, moderate susceptibility, high susceptibility, and very high susceptibility using the natural breakpoint method. Table 6.5 indicates the distribution of each susceptibility class in the susceptibility zoning map. In the prediction results of the GCN model, the percentage of very low susceptibility zone is 63.17%, the percentage of low susceptibility zone is 29.80%, the percentage of medium susceptibility zone is 3.96%, the percentage of high susceptibility zone is 2.69%, and the percentage of very high susceptibility zone is 0.36%. The susceptibility zoning map

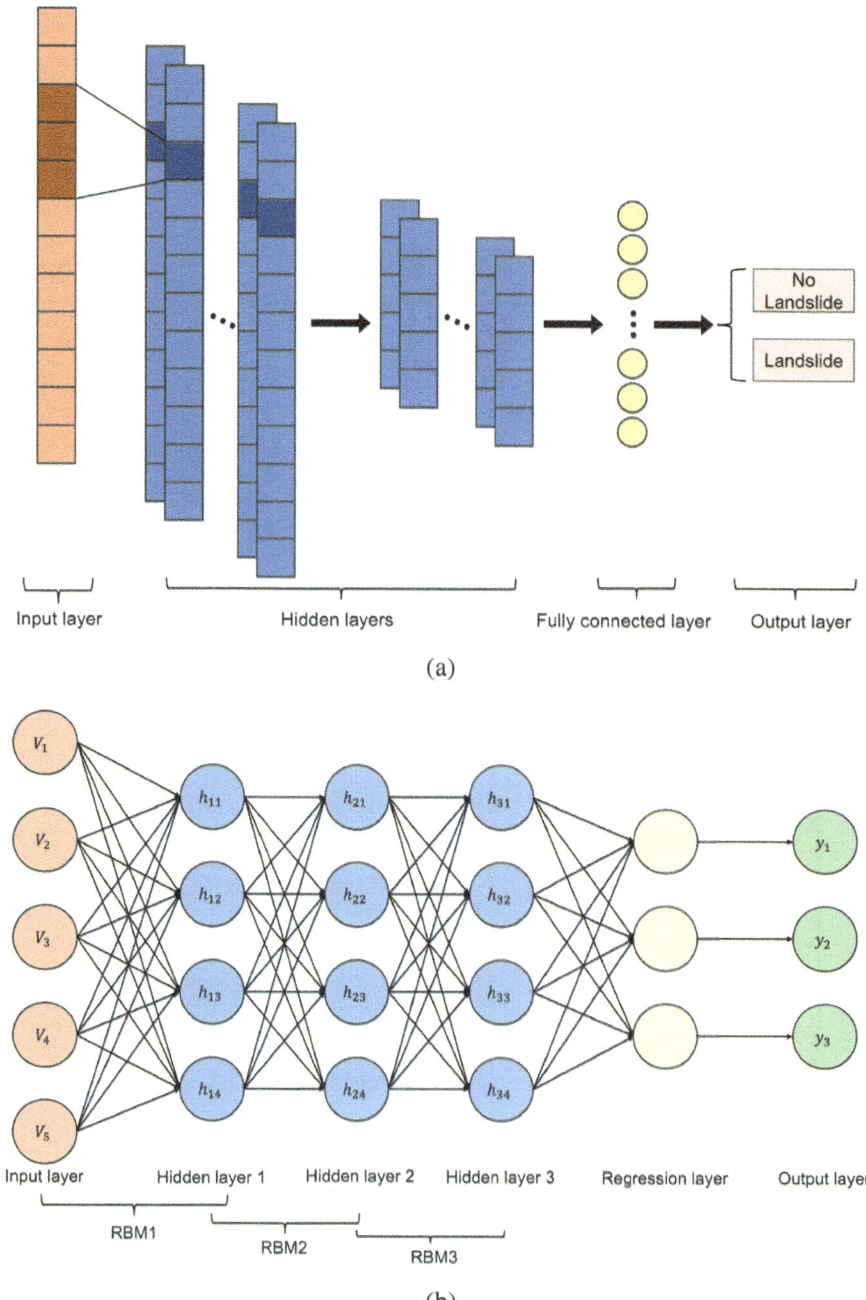


**Fig. 6.16** **a** CNN-1D model structure, modified from Wang et al. (2019) and **b** DBN model structure, modified from Wang et al. (2020a, 2020b)

**Table 6.4** Comparison of model performances

| Model | Accuracy | Precision | Recall | F1-score |
|---|---|---|---|---|
| GCN | **0.93** | **0.95** | 0.91 | **0.93** |
| CNN-1D | 0.91 | 0.90 | **0.93** | 0.91 |
| DBN | 0.91 | 0.90 | 0.92 | 0.91 |

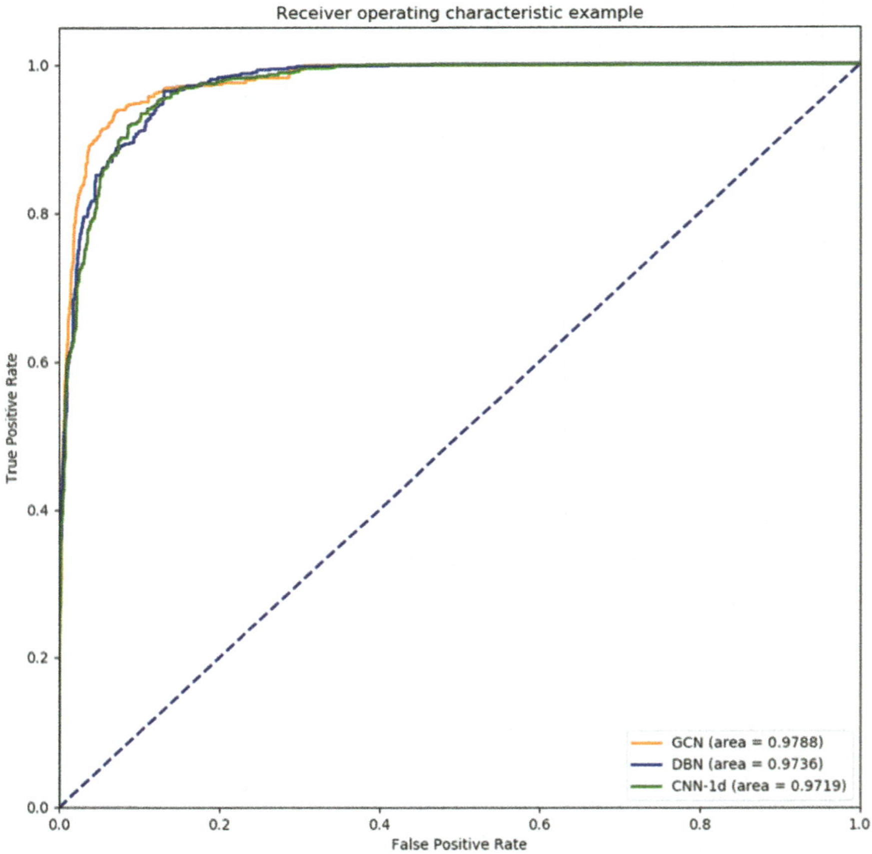


**Fig. 6.17** ROC curves of the GCN, DBN and CNN-1D models

is shown in Fig. 6.18. In the prediction results of the CNN-1D model, the distribution is very close to that of the GCN model. However, it is different in DBN model. The percentage of very low susceptibility zone is 76.10%, much higher than the previous two models. The percentages of moderate and high susceptibility zone are also higher. In contrast, the percentage of low susceptibility zone is 10.38%, which is much lower.

**Table 6.5** Statistical table of susceptibility assessment grade results

| Susceptibility | GCN | | CNN | | DBN | |
|---|---|---|---|---|---|---|
| | Pixels | Percentage (%) | Pixels | Percentage (%) | Pixels | Percentage (%) |
| Very low | 16,506,684 | 63.17 | 16,589,268 | 63.48 | 19,886,112 | 76.10 |
| Low | 7,787,628 | 29.80 | 7,475,184 | 28.60 | 2,712,600 | 10.38 |
| Moderate | 1,034,928 | 3.96 | 1,184,724 | 4.53 | 1,964,412 | 7.51 |
| High | 704,412 | 2.69 | 766,476 | 2.93 | 1,492,596 | 5.71 |
| Very high | 96,048 | 0.36 | 114,048 | 0.43 | 73,980 | 0.28 |

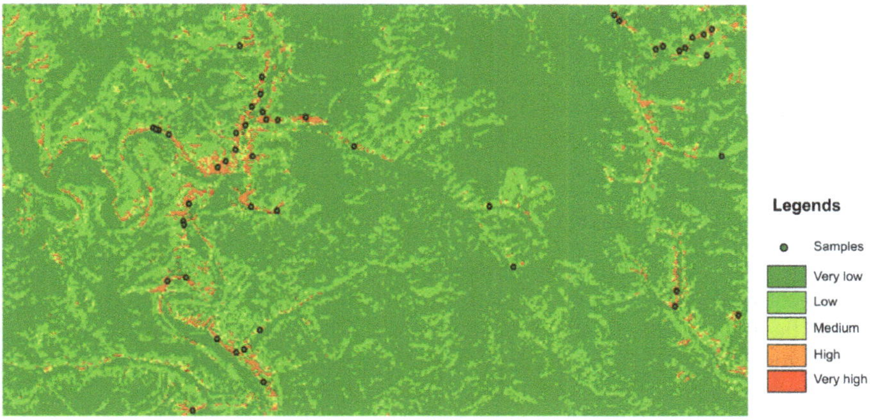


**Fig. 6.18** Landslide susceptibility zoning map

Overall, all the predicted results indicate same distribution that pixels in different susceptibility classes can be ordered as very low susceptibility > low susceptibility > moderate susceptibility > high susceptibility > very high susceptibility, which is in line with the actual situation.

## 6.5 Summary

In this chapter, a DL-based landslide susceptibility assessing model is proposed to generate the pixel-based landslide susceptibility zoning map for Yanbian Korean Autonomous Prefecture, China. Landslide related factors are selected as the input features to construct the experimental dataset. To tackle the problems of traditional methods of geodisaster susceptibility assessment, such as high dependance on prior knowledge and weak generalization ability, GCN module is introduced to the DL model. The model performance is compared with other DL-based models. Finally, landslide susceptibility zoning map with 5 susceptibility classes is generated for the study site. The main findings are as follow:

(1) Slope angle, elevation, slope aspect, distance to road, fault, land use, stratigraphic lithology and distance to water are selected as input features of the mode. The feature maps are resampled to 5 × 5 m pixels. A total of 33,218 landslide pixels and 33,218 non-landslide pixels are extracted, which are converted to 2D forms to construct the landslide experimental dataset.
(2) A landslide susceptibility assessing model based on GCN module is constructed. The landslide susceptibility of the study area is classified into five classes of very low susceptibility, low susceptibility, moderate susceptibility, high susceptibility and very high susceptibility by using natural breakpoint method to generate the final landslide susceptibility zoning map.
(3) The proposed model is compared with CNN-1D and DBN models. The prediction accuracy of GCN model is the highest of 95%. The ROC curves of the three models are plotted and the AUC value of GCN model is also high than the rest two models. This proved the GCN model can accurately predict landslide susceptibility in this dataset.

The research can be potentially improved from the following aspects:

(1) The area of landslide in the study site is small and there is limited representativeness of the experimental samples. More landslide data should be included in the future work. In addition, the number of features used is inadequate. The factors leading to landslide can be numerous, so the subsequent methods can be used to select features with greater relevance to landslide.
(2) The data is input only in the form of 2D matrix. Multiple expressions can be used for the input data. Pixel-based susceptibility assessing method will generate a large number of units with high redundancy, resulting in a slower training speed. Therefore, topographic and geomorphological units can be considered to be the assessing units.
(3) Only the landslide in Yanbian Korean Autonomous Prefecture is selected as the object geodisaster for susceptibility analysis. There are other types of geodisaster like ground fissure in this region. Subsequent analysis of multiple geodisasters can be carried out to provide more scientific suggestions for geodisaster prevention and control.

## Codes

### [Python3.7] CNN and GCN

```
import torch
import torch.nn as nn
import torch.nn.functional as F
import numpy as np
class GCNChannel(nn.Module):
  def __init__(self, channels):
```

```
        super(GCNChannel, self).__init__()
        self.input = nn.Sequential(
            nn.Conv2d(channels, channels, kernel_size = 3, stride = 1,
padding = 1),
            nn.BatchNorm2d(channels),
            nn.ReLU(inplace = True)
        )
        self.fc1 = nn.Conv1d(channels, channels, kernel_size = 1, bias
= False)
        self.fc2 = nn.Conv1d(channels, channels, kernel_size = 1, bias
= False)
    def pre(self, x):
        b, c, h, w = x.size()
        x = x.view(b, c, -1).permute(0, 2, 1)#[10,144,8]
        x = x.view(b, 1, h * w, c)#[10,1,144,8]
        return x
    def normalize(self, A):
        b, c, im = A.size()
        out = np.array([])
        for i in range(b):
            # A = A = I
            A1 = A[i].to(device = "cpu")
            I = torch.eye(c, im)
            # A1 = A1 + I
            # degree matrix
            d = A1.sum(1)
            # D = D^-1/2
            D = torch.diag(torch.pow(d, -0.5))
            new_A = D.mm(A1).mm(D).detach().numpy()
            out = np.append(out, new_A)
        out = out.reshape(b, c, im)
        normalize_A = torch.from_numpy(out)
        normalize_A = normalize_A.type(torch.FloatTensor)
        return normalize_A
    def forward(self, x):
        b, c, h1, w1 = x.size()
        x = self.pre(x)
        A = torch.ones((b, h1*w1, h1*w1))
        A = self.normalize(A)
        x = x.view(b, -1, c)
        x = F.relu(self.fc1(A.bmm(x).permute(0, 2, 1))).permute(0, 2,
1)
        x = self.fc2(A.bmm(x).permute(0, 2, 1))
        out = x.view(b, c, h1, w1)
        out = F.interpolate(out, size = (h1, w1), mode = 'bilinear',
align_corners = True)
        return out
class GCNSpatial(nn.Module):
    def __init__(self, channels):
        super(GCNSpatial, self).__init__()
        self.fc1 = nn.Conv1d(channels, channels, kernel_size = 1, bias
= False)
        self.fc2 = nn.Conv1d(channels, channels, kernel_size = 1, bias
= False)
```

```
    def normalize(self, A):
        b, c, im = A.size()
        out = np.array([])
        for i in range(b):
            A1 = A[i].to(device = "cpu")
            # I = torch.eye(c, im)
            # A1 = A1 + I
            # degree matrix
            d = A1.sum(1)
            # D = D^-1/2
            D = torch.diag(torch.pow(d, -0.5))
            new_A = D.mm(A1).mm(D).detach().numpy()
            out = np.append(out, new_A)
        out = out.reshape(b, c, im)
        normalize_A = torch.from_numpy(out)
        normalize_A = normalize_A.type(torch.FloatTensor)
        return normalize_A
    def forward(self, x):
        b, c, h, w = x.size()
        A = torch.ones((b,c,c))
        A = self.normalize(A)
        x = x.view(b, c, -1)
        x = F.relu(self.fc1(A.bmm(x)))
        x = self.fc2(A.bmm(x))
        out = x.view(b, c, h, w)
        return out
class CNN(nn.Module):
    def __init__(self):
        super(CNN, self).__init__()
        self.conv1 = nn.Sequential( #input shape (1,12,12)
            nn.Conv2d(in_channels = 1, #input height
                    out_channels = 8, #n_filter
                    kernel_size = 3, #filter size
                    stride = 1, #filter step
                    padding = 1 #con2d
                   ), #output shape (16,12,12)
            nn.ReLU(),
        )
        self.depth_conv = nn.Sequential(
            nn.Conv2d(in_channels = 8, out_channels = 8, kernel_size =
3, stride = 1, padding = 1, groups = 8),
            nn.BatchNorm2d(8),
            nn.ReLU(inplace = True)
        )
        self.point_conv = nn.Sequential(
            nn.Conv2d(in_channels = 8, out_channels = 8, kernel_size =
1),
            nn.BatchNorm2d(8),
            nn.ReLU(inplace = True)
        )
        out_channels = 8
        num_classes = 2
        self.gcn_c = GCNChannel(out_channels)
        self.gcn_s = GCNSpatial(out_channels)
```

```
        self.conv2 = nn.Sequential(
            nn.Conv2d(out_channels, 4, kernel_size = 3, stride = 1,
    padding = 1),
          nn.BatchNorm2d(4),
          nn.ReLU(),
          nn.MaxPool2d(kernel_size = 2),
        )
        self.output = nn.Sequential(
          nn.Linear(4*6*6, 6*6),
          nn.Dropout(p = 0.5),
          nn.Linear(36,2),
        )
      def forward(self, x):
        b,c,h,w = x.size()
        x = self.conv1(x)
        x_s = self.depth_conv(x)
        x_c = self.point_conv(x)
        x_s = self.gcn_s(x_s)
        x_c = self.gcn_c(x_c)
        x = x_c + x_s
        x = self.conv2(x)
        x = x.view(b,-1)
        output = self.output(x)
        return output
```

## References


Bruna, J., Zaremba, W., Szlam, A., & LeCun, Y. (2013). *Spectral networks and locally connected networks on graphs*. arXiv preprint, arXiv:1312.6203.

Chen, W., Xie, X., Peng, J., Wang, J., Duan, Z., & Hong, H. (2017). GIS-based landslide susceptibility modelling: A comparative assessment of kernel logistic regression, Naïve-Bayes tree, and alternating decision tree models. *Geomatics, Natural Hazards and Risk, 8*(2), 950–973.

Department of Natural Resources of Jilin Province. (2019). *Jilin Province geological disater report in 2019*. Retrieved from http://www.yb983.com/p/97885.html

Ermini, L., Catani, F., & Casagli, N. (2005). Artificial neural networks applied to landslide susceptibility assessment. *Geomorphology, 66*, 327–343.

Feng, H., Zhou, A., Yu, J., Tang, X., Zheng, J., Chen, X., & You, S. (2016). A comparative study on plum-rain-triggered landslide susceptibility assessment models in West Zhejiang Province. *Earth Science, 41*(3), 403–415.

Fu, W. (2008). Landslide hazard evaluation based on GIS and SVM. *Scientia Geographica Sinica, 28*(6), 838–841.

Gori, M., Monfardini, G., & Scarselli, F. (2005). A new model for learning in graph domains. *Proceedings of the IEEE International Joint Conference on Neural Networks, 2*, 729–734.

Grozavu, A. (2021). Mapping landslide susceptibility at national scale by spatial multi-criteria evaluation. *Geomatics, Natural Hazards and Risk, 12*(1), 1127–1152.

Hong, H., Pradhan, B., Jebur, M. N., Bui, D. T., Xu, C., & Akgun, A. (2016). Spatial prediction of landslide hazard at the Luxi area (China) using support vector machines. *Environmental Earth Sciences, 75*(1), 1–14.

Howard, A. G., Zhu, M., Chen, B., Kalenichenko, D., Wang, W., Weyand, T., et al. (2017). *Mobilenets: Efficient convolutional neural networks for mobile vision applications*. arXiv preprint, arXiv:1704.04861.

Hu, T. (2020). *Study on geological hazard assessment of Sinan County, Guizhou Province*. PhD. Thesis, China University of Geosciences (Wuhan).

Huang, W., Ding, M., Wang, D., Jiang, L., & Li, Z. (2022). Evaluation of landslide susceptibility based on layer adaptive weighted convolutional neural network model along Sichuan-Tibet Traffic Corridor. *Earth Science, 47*(6), 2015–2030.

Jing, W., Li, X., & Yang, J. (2022). Evaluation of landslide susceptibility in Linzhi District, Changdu based on deep belief network. *Railway Standard Design, 66*(9), 7–14.

Kipf, T. N., & Welling, M. (2016). *Semi-supervised classification with graph convolutional networks*. arXiv preprint, arXiv:1609.02907.

LeCun, Y., Bottou, L., Bengio, Y., & Haffner, P. (1998). Gradient-based learning applied to document recognition. *Proceedings of the IEEE, 86*(11), 2278–2324.

Li, W., Fang, Z., & Wang, Y. (2022). Stacking ensemble of deep learning methods for landslide susceptibility mapping in the Three Gorges Reservoir area, China. *Stochastic Environmental Research and Risk Assessment, 36*(8), 2207–2228.

Liang, Z., Wang, C., Han, S., Ullah Jan Khan, K., & Liu, Y. (2020). Classification and susceptibility assessment of debris flow based on a semi-quantitative method combination of the fuzzy C-means algorithm, factor analysis and efficacy coefficient. *Natural Hazards and Earth System Sciences*, 20(5), 1287–1304.

Mallick, J., Singh, R. K., AlAwadh, M. A., Islam, S., Khan, R. A., & Qureshi, M. N. (2018). GIS-based landslide susceptibility evaluation using fuzzy-AHP multi-criteria decision-making techniques in the Abha Watershed, Saudi Arabia. *Environmental Earth Sciences, 77*(7), 1–25.

Ministry of Natural Resources of the People's Republic of China. (2022a). *Last year, 905 geodisasters were predicted nationwide, avoiding direct economic losses of 1.35 billion RMB.*

Ministry of Natural Resources of the People's Republic of China. (2022b). *National Geodisaster Situation in 2021 and Trend Prediction in 2022b.*

Oh, H.-J., Kadavi, P. R., Lee, C.-W., & Lee, S. (2018). Evaluation of landslide susceptibility mapping by evidential belief function, logistic regression and support vector machine models. *Geomatics, Natural Hazards and Risk, 9*(1), 1053–1070.

Oh, H.-J., & Lee, S. (2017). Shallow landslide susceptibility modeling using the data mining models artificial neural network and boosted tree. *Applied Sciences, 7*(10), 1000.

Pham, T. B., Bui, T. D., Prakash, I., & Dholakia, B. M. (2016). Rotation forest fuzzy rule-based classifier ensemble for spatial prediction of landslides using GIS. *Natural Hazards, 83*(1), 97–127.

Reichenbach, P., Rossi, M., Malamud, B. D., Mihir, M., & Guzzetti, F. (2018). A review of statistically-based landslide susceptibility models. *Earth-Science Reviews, 180*, 61–91.

Sun, W., Tan, C., Wang, J., Wu, S., & Zhang, C. (2008). Geohazard susceptibility evaluation of Qianyang County, Baoji area, Shaanxi, China. *Geological Bulletin of China, 27*(11), 1847–1853.

Tang, C., & Ma, G. (2015). Small Regional Geohazards Susceptibility Mapping Based on Geomorphic Unit. *Scientia Geographica Sinica, 35*(1), 91–98.

Wang, Z., Yi, F., & Chen, T. (2012). Geo-hazard susceptibility evaluation of Mianyang City based on fuzzy comprehensive evaluation. *Science % Technology Review*, 16, 53–60.

Wang, H., Zhang, L., Luo, H., He, J., & Cheung, R. W. (2021). AI-powered landslide susceptibility assessment in Hong Kong. *Engineering Geology, 288*, 106103.

Wang, W., He, Z., Han, Z., Li, Y., Dou, J., & Huang, J. (2020a). Mapping the susceptibility to landslides based on the deep belief network: A case study in Sichuan Province, China. *Natural Hazards, 103*(3), 3239–3261.

Wang, Y., Fang, Z., Wang, M., Peng, L., & Hong, H. (2020b). Comparative study of landslide susceptibility mapping with different recurrent neural networks. *Computers and Geosciences*, 138, 104445.

Wang, Y., Fang, Z., & Hong, H. (2019). Comparison of convolutional neural networks for landslide susceptibility mapping in Yanshan County, China. *Science of the Total Environment, 666*, 975–993.

Wu, B., Liang, X., Zhang, S., & Xu, R. (2020a). Advances and applications in graph neural network. *Chinese Journal of Computers, 45*(1), 35–68.

Wu, Z., Pan, S., Chen, F., Long, G., Zhang, C., & Philip, S. Y. (2020b). A comprehensive survey on graph neural networks. *IEEE Transactions on Neural Networks and Learning Systems, 32*(1), 4–24.

Yang, S. (2016). *Research on evaluation method of geological disaster susceptibility in southeast coastal areas: Taking Fujian Province as an example.* Master Thesis, Beijing Jiaotong University.

Yi, Y., Zhang, Z., Zhang, W., Jia, H., & Zhang, J. (2020). Landslide susceptibility mapping using multiscale sampling strategy and convolutional neural network: A case study in Jiuzhaigou region. *CATENA, 195*, 104851.

Zhang, Y.-L., Yan, D., & Heng, P.-A. (2005). Quantitative analysis of the relationship of biology species using Pearson correlation coefficient. *Computer Engineering and Applications, 33*, 79–82.

Zhang, Y., Zhang, C., & Zhang, L. (1993). Analytic hierarchy processing and computation of the aggregative extent of disaster damage on China's geological disaster system. *Bulletin of the Chinese Academy of Geological Sciences, 28*, 139–154.

Zhu, Q. (2020). *Assessment of geological hazard susceptibility in Baqiao District based on RF and SVM models.* Master Thesis, Xi'an University of Science and Technology.

# Chapter 7
# Deep Learning Based Intelligent Recognition of Ground Fissures

**Abstract** Manual detection and labelling of ground fissures are inefficient and of high error rate. To develop the automatic method for ground fissure location, this chapter proposes a ground fissure detection method based on deep learning (DL) and multi-scale map convolution. Based on U-Net semantic segmentation model, a multi-scale global reasoning module (MGRB) is added in the central part of the network. The MGRB is composed of multi-scale pooling operator and graph reasoning module, with multiple receptive fields to learn the features of ground fissures in different sizes. The module contains four branches to obtain the context features in a larger range and each branch contains a graph reasoning module. The original features are projected into the node space and are thereby convoluted to extract the global node features, which are finally fused with the original features and back projected into the original feature space. The proposed model is proved to be more feasible and effective in the ground fissure segmentation task comparing with traditional digital image processing methods and other DL-based methods.

## 7.1 Introduction

### *7.1.1 Research Background and Significance*

The formation of ground fissures is caused by the regional tectonic background and tectonic stress field. It is the manifestation of crustal neotectonics. Shanxi fault basin of China has a large population and developed agriculture, which has been the core economy region in China since ancient time (Meng, 2011). Most of the important engineering activities are located in the fault basin, including large industrial bases, the west to east gas transmission project and the high-speed railway lines. However, this region has strong tectonic activities and frequent geodisasters. The crustal activities started to happen frequently since the beginning of Quaternary (Jiang et al., 1997). In the middle of the twentieth century, many ground fissures were recorded in Shanxi Graben System. In 1970s, ground fissures occurred in most basins of the fault zone of Shanxi Province, and their scales were different. In 1980s, ground fissure became more serious and caused significant economic loss and threatened human life

 
W. Chen et al., *Intelligent Interpretation for Geological Disasters*,
https://doi.org/10.1007/978-981-99-5822-1_7

and property. Therefore, it is of great significance to improve the ability to manage the ground fissure.

Most ground fissures in Shanxi Province are connected with active faults, which is closely related to tectonic activities (Yang et al., 1999). There are also some ground fissures generated by the continuous scouring of rainstorms. In addition, the change of groundwater level can also lead to ground fissures, as human exploit large amount of groundwater and the stress in the underground environment becomes uneven and gradually subsides. Due to the diversity of the causes, as well as complex properties including the huge number, shape diversity, spatial distribution randomness, it is difficult to identify the ground fissures artificially. In recent years, with the rapid development of machine learning, deep learning and computer vision, artificial recognition of ground fissures is no longer applicable to current projects. The manual segmentation method is not only time-consuming and laborious, but also cannot meet the requirement of all-time available. Missing and wrong detection easily occur in the dense area of ground fissures. Therefore, how to achieve the automation and intelligence location of ground fissure has become an urgent problem to be solved.

### *7.1.2 Research Overview*

In the last century, the labelling of ground fissure image mainly depended on manual work. In the case with a large number of samples, this method is time-consuming and easily to cause missing and wrong detections, so the segmentation accuracy was not ideal. Digital image processing has been developing towards faster processing speed, higher efficiency and deeper image semantic information since its invention. Machine learning also has been developed with a large number of advanced algorithms. These developments create conditions for image-based lossless segmentation technology, which has become a research hotspot of ground fissure extraction (Ma, 2019). Two dominant methods for ground fissure extraction are based on digital image processing and deep learning.

**Fissure Extraction Based on Digital Image Processing**

Li et al. (1991) extracted the edge of the image by the classical Sobel edge detection operator and achieved the detection of ground fissure by removing the noise whose edge perimeter is less than the given value. However, this method did not perform well due to the complex fissure topology and uneven noise size and shape. Then, the method combining Sobel operator, Kittler automatic threshold selection and processing operation sequence was proposed for the extraction of fine ground fissures in asphalt pavement. As the ground fissure always indicates the linear shape structure, Gao et al. (2003) limited the size of the minimum circumscribed matrix and then set discrimination threshold to identify ground fissures. However, performance of this method can be significantly affected by image noise. Subirats et al. (2006) proposed a two-dimensional discrete wavelet transform to process the image and then calculated the maximum value of the wavelet coefficient to determine if there is ground fissure in the image. Considering the ground fissures always showed discontinuity, Liu et al. (2008) proposed an automatic fissure extraction method for complex pavement image based on region growth algorithm. But this method is of high computational complexity. If the image noise and gray value distribution are uneven, it will lead to holes and over segmentation. Xu et al. (2013) proposed a pavement fissure extraction method based on the image saliency. Based on the local brightness, edge characteristics and continuity characteristics of fissures, the fine-scale local saliency can be enhanced. Then the adaptive threshold segmentation is used to segment the fissures. Kaddah et al. (2020) proposed an improved voting fissure extraction algorithm based on the optical and geometric characteristics of road fissures, which can extract fine fissures and fissures with low contrast. He et al. (2020) proposed an adaptive threshold segmentation algorithm based on Otsu algorithm to extract fine fissures with complex background considering the invariance of Hessian matrix rotation to enhance the edge of the image. Partial fissures were then combined according to the growth direction.

**Fissure Extraction Based on Depth Learning**

Zhang et al. (2016) firstly applied convolution neural network (CNN) to detect road fissures with an overall accuracy of 86.96%. Zhang et al. (2017a) proposed a crack segmentation network CrackNet. Filters with different directions, lengths and widths were used to generate feature maps as the input. Pooling layer was removed and all hidden layers consisted of convolution layers and fully connected layers. The accuracy can be improved, but model requires too many parameters. Wang et al. (2018) proposed the Crack-FCN model which applied the full convolution network to fissure segmentation. It can extract more complete fissures and learn deeper features to reduce influence of noise. However, this method is insensitive to the image details in the upsampling layer, which might lead to misclassification. Weng et al. (2019) also used the end-to-end full convolution network to achieve the automatic segmentation of ground fissures. Zou et al. (2018) proposed the DeepCrack model which is an end-to-end deep convolution network. It can restore more details of the image after learning the multidimensional features in deep layers and is insensitive to noise, which allows the model to restore the fissure image with low contrast. Yang (2019) used SPPNet network by the combination of Deeplab and PSPNet for tunnel fissure segmentation. High-level semantic features are extracted from the image and collected in six different levels to obtain the multi-scale information of the image, which can further reduce the loss of feature information and improve the segmentation accuracy. Choi and Cha (2019) improved the SDDNet network for real-time concrete fissure segmentation with DenSep module and ASPP module. Although this method requires obvious fissure features, training parameters required is significantly reduced, and the training speed can be much faster than other networks. Zhang (2020) used an improved dense connected convolution network called DenseNet to segment tunnel fissures, which can reduce the influence of environmental factors and achieve the fuzzy segmentation at image level and pixel level. Li et al. (2019)

developed the DBCC model for bridge fissure extraction with sliding window algorithm. Image pyramid and the region of interest were fused to improve the temporal performance of the model.

### 7.1.3 Research Object and Contents

According to the literatures, the main issues in the automatic fissure detection in current stage include: (1) Labelling of fissure samples; (2) The accuracy of fissure segmentation is low for fissures with complex background; (3) The accuracy of fissure segmentation is low for fissures with complex topology. To tackle these issues, the main contents of this paper are as follows:

(1) Considering the lack of labeled sample of ground fissures, a dataset of is constructed by using the remote sensing images of Shanxi Province in this study. The images are pre-processed by image color enhancement, cropping, rotation. Data augmentation algorithms including Mixup data augmentation, cutout data augmentation and mosaic data augmentation are used to obtain more labeled data and generate the final dataset.
(2) U-net based network with multi-scale graph convolution feature is proposed for ground fissure segmentation. Graph convolution network (GCN) introduced to learn graph data with irregular structures. The overall framework of the segmentation method includes three parts: (a) The first part is the phase for the feature extraction with batch normalization module. (b) The second part contains a multi-scale global reasoning module MGRM. The input image is processed through four different paths to obtain the output image, and each path contains a graph reasoning module to obtain the global feature information of the input image. (c) The final part is the image restoration phase that uses transpose convolution to replace the upsampling function.
(3) The performance of the proposed model is tested by comparing the segmentation results with evaluation metrics and different ground fissure segmentation models.

The content structure of this chapter is arranged as follows: Sect. 7.2 introduces technologies used in this study, including U-Net network and GCN. Section 7.3 introduces the acquisition and pre-processing of datasets, and data augmentation methods used in this study. Section 7.4 describes the ground fissure segmentation framework based on the multi-scale convolution feature map. Section 7.5 is the results and analysis of the experiment. The feasibility and efficiency of the proposed model are proved by comparing the performance with other models. Section 7.6 summarizes the research and proposes improvement suggestions for the problems of current research method.

## 7.2 Related Principles and Technologies

### 7.2.1 U-Net

U-net (Ronneberger et al., 2015) is originally published at the 2015 MICCAI conference which has a shape similar to the letter "U" with the layer hopping connection and the fine FCN in decoding stage. Because of its excellent performance in image segmentation, it has been widely used in many fields, including the satellite image segmentation and industrial product defect detection. The structure of U-Net is indicated in Fig. 7.1.

The left part of U-Net is the feature extraction phase, also called the encoder. Image features are extracted by using two 3 × 3 convolutions, followed by a ReLU activation function. The length and width of the input image will decrease by 2 pixels in each convolution operation. As shown in Fig. 7.1, the 572 × 572 image size becomes 568 × 568 after two convolutions. After that, the maximum pooling layer with 2 × 2 kernel is used to decrease the size of the feature map to generate an image that is of half width and height of the original image. In Fig. 7.1, the 568 × 568 size image is downsampled to 284 × 284 in the first downsampling. The encoder performs the abovementioned operations for four times, and the size of the image finally decreases to one-sixteenth of the original image.

The right part is the upsampling phase, also called the decoder. In this part, the encoded image is expanded by deconvolution with a stride of 2 to generate an image that is of twice width and height of the encoded image. In Fig. 7.1, the image from the fifth floor with size of 28 × 28 is upsampled to the image with size of 56 × 56. Then, the feature map obtained from corresponding layers in the feature extraction stage is cropped, copied, and spliced with the feature map obtained by upsampling.

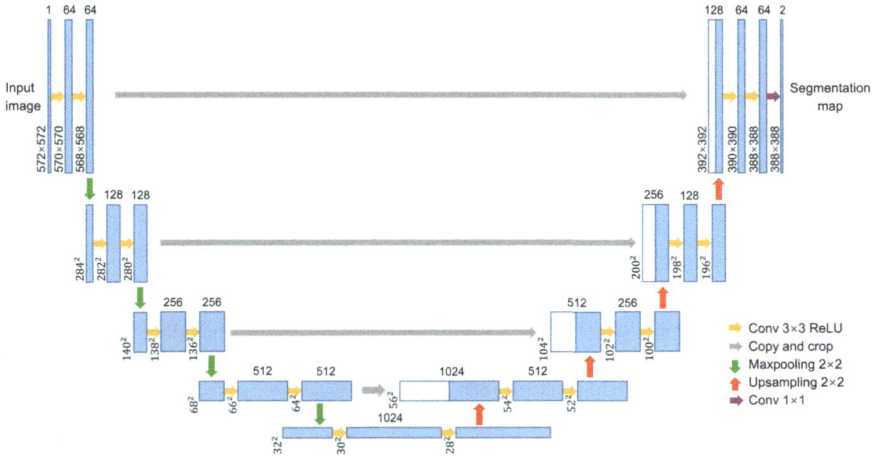


**Fig. 7.1** U-net architecture

Two $3 \times 3$ convolutions are used for feature extraction, which are activated by ReLU activation functions. The decoder also performs the abovementioned operations for four times.

In the image output phase, sigmoid or softmax activation function is used to output the segmented image. U-Net contains layer hopping connection to fully integrate high-level semantic information and low-level location information to generate more accurate results.

### 7.2.2 Graph Convolution Network

The traditional CNN is limited to modeling the data with very regular Euclidean spatial topology. Image is a two-dimensional pixel-based data. Regarding a pixel in the image as the central node, the number of neighbored nodes is constant. CNN can extract the local features by learning these translation invariance and local connectivity properties of the Euclidean data. However, in practice, the data structure is usually irregular, such as topological data and graph data. Chemical molecular structure, knowledge graph and social network are all such irregular data structures. The structures around nodes are different and the number of neighbor nodes can be unfixed, which is consistent with the property that infinite dimensional normed linear space has a bounded set but not dense. This feature makes the traditional RNN and CNN not effective in these cases. To process graph data, graph convolution neural network (GNN) introduced to apply the convolution in graph data. GCN (Kipf & Welling, 2016) is one of the most popular GNN.

GCN was proposed at the 2017 ICLR conference. The network can be directly used to process graph data. In addition, the model scale of GCN can increase linearly with the increase of the number of edges in the graph. Supposing that the graph data includes $N$ nodes, defined as $G = (V, E)$, where $v_i \in V$ indicates node, $e_{i,j} \in E$ represents the edge between node $i$ and node $j$. The relationship between nodes is represented by adjacency matrix $A \in R^{N \times N}$. If each node contains $D$-dimensional features, the features of these nodes will form an $N \times D$-dimensional matrix $X$, and $X \in R^{N \times D}$. In essence, GCN is a local first-order approximation of spectral convolution, and its main idea is to use the Laplacian change of the frequency spectrum for graph convolution. The equation of Laplace matrix is as follow:

$$L = D - A \tag{7.1}$$

where $D$ is the degree matrix of the graph and $A$ is the adjacency matrix. Laplacian matrix is both a real symmetric matrix and a positive semi-definite matrix. Here, the Laplacian matrix is spectral decomposed:

$$L = I_N - D^{-\frac{1}{2}} A D^{-\frac{1}{2}} = U \Lambda U^T \tag{7.2}$$

where $U$ is the orthogonal matrix composed of eigenvectors of normalized Laplace matrix, $\Lambda$ is the diagonal matrix of corresponding eigenvalues, $U^T$ is the transpose of the matrix $U$. And the spectral convolution on the graph can be defined as the product of signal $x \in R^N$ and filter $g_\theta = diag(\theta), \theta \in R^N$ in Fourier domain:

$$xG_g = F^{-1}[F(x) \cdot F(g)] \tag{7.3}$$

where $x$ represents the features of the node. $G_g$ represents convolution kernel in frequency domain. $F()$ indicates Fourier transform. $F^{-1}()$ indicates inverse Fourier transform. If $U$ is taken as the basis function of graph Fourier transform, Eq. (7.3) is equivalent to Eq. (7.4):

$$xG_g = U\left(U^T x \cdot U^T g\right) \tag{7.4}$$

When $g_\theta = diag\left(U^T g\right)$, then Eq. (7.4) can be expressed as Eq. (7.5):

$$xG_g = U_{g_\theta} U^T x \tag{7.5}$$

However, for the spectral convolution of graphs, the computational complexity of matrix multiplication using the normalized Laplace matrix $U$ is $\mathcal{O}\left(N^2\right)$. For the graph with very complex topological structure, it is very costly to calculate the eigen decomposition of Laplace matrix $L$. Thus, the K-order truncation of Chebyshev polynomial in Eq. (7.6) is used to obtain the approximation of $G_g$, so Eq. (7.5) is approximate to Eq. (7.7):

$$T_k(x) = 2xT_{k-1}(x) - T_{k-2}(x) \tag{7.6}$$

$$xG_g \approx \sum_{k=0}^{K} \theta_k T_k\left(\hat{L}\right)x \approx \theta_0 x + \theta_1 (L - I_N)x = \theta_0 x - \theta_1 D^{-\frac{1}{2}} A D^{-\frac{1}{2}} \tag{7.7}$$

Equation (7.7) contains two parameters $\theta_0$ and $\theta_1$. In the practical calculation process, overfitting can be avoided by restricting the parameter, which can also simplify the operations. Setting $\theta = \theta_0 = -\theta_1$ to get Eq. (7.8):

$$xG_g \approx \theta\left(I_N + D^{-\frac{1}{2}} A D^{-\frac{1}{2}}\right)x \tag{7.8}$$

where $\theta$ is the learnable weight. However, eigenvalue range of $I_N + D^{-\frac{1}{2}} A D^{-\frac{1}{2}}$ is [0, 2]. When the network is very deep, the gradient explosion or disappearance may occur if the operation is repeated. Therefore, a renormalization trick is adopted:

$$I_N + D^{-\frac{1}{2}} A D^{-\frac{1}{2}} \rightarrow \tilde{D}^{-\frac{1}{2}} \tilde{A} \tilde{D}^{-\frac{1}{2}} \tag{7.9}$$

where $\tilde{A} = A + I_N$ is normalized adjacency matrix. $\tilde{D} = \sum_j \tilde{A}_{ij}$ is the normalized degree matrix.

After the above derivation, the network propagation formula of GCN is obtained:

$$H^{(l+1)} = \delta\left(\tilde{D}^{-\frac{1}{2}}\tilde{A}\tilde{D}^{\frac{1}{2}}H^{(l)}W^{(l)}\right) \tag{7.10}$$

where the initial input is $H^{(0)} = X$, $H^{(l)} \in R^{N\times D}$ is the feature of node in layer $l$, $W^{(l)} \in R^{D\times D}$ is the parameter to be trained in layer $l$, $\delta()$ represents a nonlinear activation function.

## 7.3 Data Acquisition and Processing

### *7.3.1 Data Source*

In this study, remote sensing multispectral images of Bagong Town in Jincheng City, China are used to prepare the ground fissure dataset. Zongheng CW-007 vertical take-off and landing fixed-wing UAV and Changguangyu MS600 multi-spectral camera were used to acquire the ground fissure images from the airspace at a height of 350 m. The survey area was about 16 km$^2$, and the images include R, G, B three bands. Most parts of the study area are covered with vegetation, and the terrain is undulating. The southwest regions are mainly villages and croplands, while the northeast regions are mainly mountains. To ensure the quality of the collected images, the following requirements should be satisfied:

(1) The image should be of enough sharpness, without cloud, shadow and large-area reflection;
(2) The tilt and rotation angle of the images should not exceed 12°;
(3) The UAV should keep a stable flight height, which should not exceed 30 m above the ground;
(4) In the course direction, the coverage of the imaging boundary should exceed the image area by at least two imaging baselines, and by at least one image frame in side direction.

ArcGIS Pro is used to label the ground fissures in the collected data. As shown in Fig. 7.2, on the upper side are the original images and on the bottom side are the labeled images.

Due to the large size of remote sensing images and corresponding vector files, the images should be cropped to cater for the requirement of specific GCN. ArcGIS Pro is used to subset the images. In total, 3306 cropped images with a resolution of 256 × 256 are obtained. As shown in Fig. 7.3, on the left side is one cropped image contains ground fissure, while on the right side is the image without ground fissure.

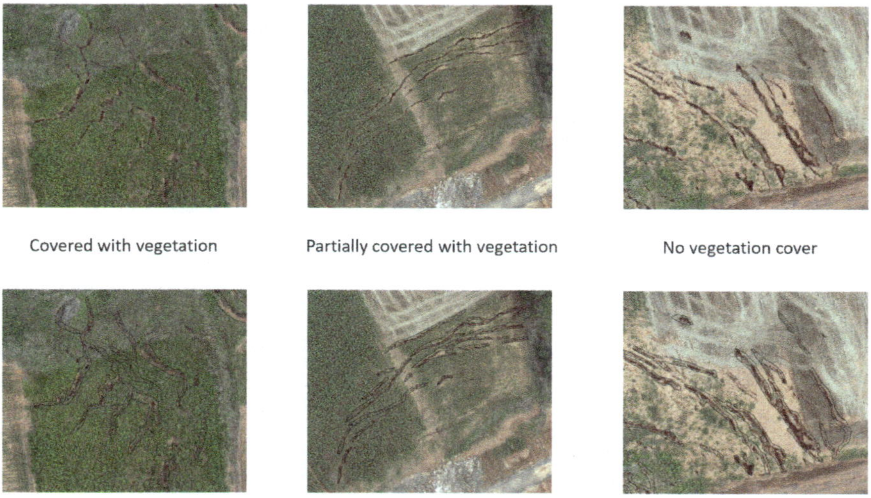


**Fig. 7.2** Original images and labeled images in ground fissure dataset

**Fig. 7.3** Cropped image in ground fissure dataset

### *7.3.2 Data Preprocessing*

The scarcity of remote sensing images and the influence of shooting angle, shooting height and lighting environment in the process of image acquisition can lead to lower quantity and quality data. In this case, the acquired images are preprocessed in construct process of the ground fissure dataset.

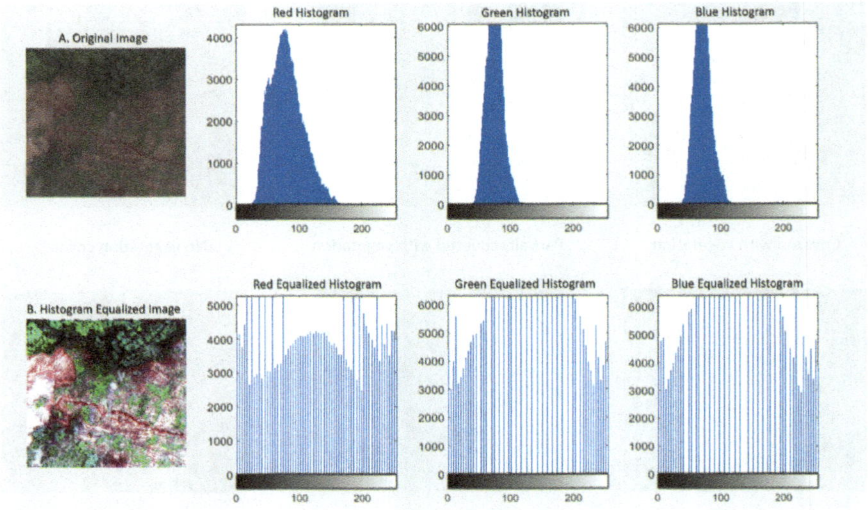


**Fig. 7.4** Histogram equalization of RGB image

### Histogram Equalization

The spectral difference caused by the sudden change of terrain makes the ground fissures appear as dark lines in remote sensing images and show linear continuity and edge characteristics. Since the gray value of the ground fissure is low in its center, the histogram was equalized to carry out the dynamic process of the gray values in each image, thereby increasing the contrast of the image.

The principle of histogram equalization is to transform the probability density distribution of known grayscale in an image and homogenize the grayscale probability of the image to obtain a new image. As shown in Fig. 7.4, the images which are too dark and too bright are equalized by histograms to be sharp, so that their contrast can be improved locally or even globally.

### Data Augmentation

In machine learning, the more complex and expressive model usually performs better on training data, but its ability to interpret unprecedented data will be significantly reduced, that is called overfitting. Too much optimization on the training data cannot be reflected on the data outside the training dataset. Larger amounts of training data are required to improve the generalization ability of the model. However, usually there are limited data available in practical cases. Therefore, in this research, the following methods are utilized to train the neural network with the small amount of training data.

(1) Augmentation of Image Color

The data augmentation of image color is mainly to adjust the saturation, contrast, brightness and some other aspects (Fig. 7.5).

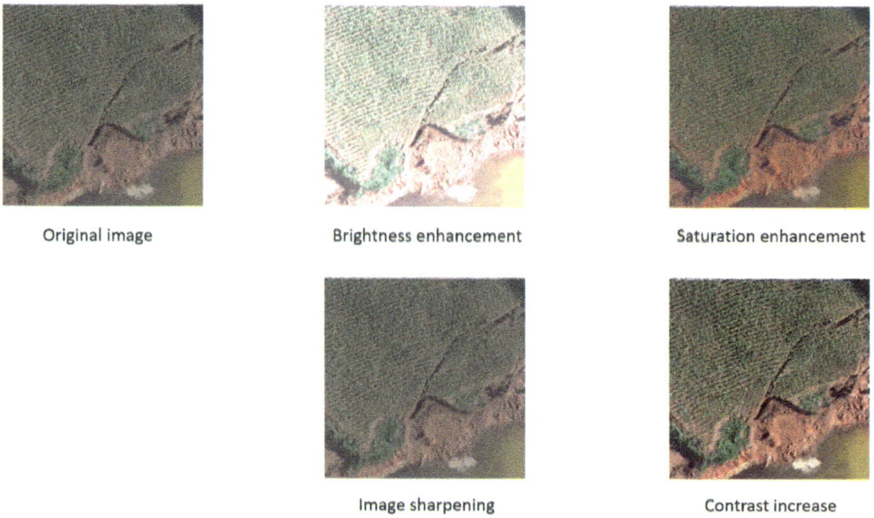


**Fig. 7.5** Data augmentation of image color

(2) Random Cropping and Rotation

The inconstant shooting height, direction and distance can result in different appearance of same object. For the limited available data, random cropping and rotation can be used to achieve the similar effect. A total of 12,092 images with the resolution of 256 × 256 are obtained after random cropping and rotation operations (Fig. 7.6).

(3) Mixup Augmentation

The performance of deep neural networks is not ideal in terms of huge memory consumption and sensitivity to adversarial examples. Mixup (Zhang et al., 2017b) is a simple solution proposed to solve these two problems. Mixup can obtain a new virtual sample for training by simply weighting and summing two random samples and corresponding labels in the training dataset. It is a mixed-class augmenting algorithm for simple images, which mixes different images in a certain proportion to expand the training dataset. Virtual samples can be constructed by Eqs. (7.11) and (7.12):

$$\tilde{x} = \lambda x_i + (1 - \lambda)x_j \tag{7.11}$$

$$\tilde{y} = \lambda y_i + (1 - \lambda)y_j \tag{7.12}$$

where $\lambda$ refers to the parameters following the Beta distribution, and $\lambda \in [0, 1]$. $(x_i, y_i)$ and $(x_j, y_j)$ are two samples randomly selected from the training dataset. $x_i$ and $x_j$ are the original input vector and $\tilde{x}$ is the mixed batch sample. $y_i$ and $y_j$ are

**Fig. 7.6** Random cropping and rotation of images

the labels in one-hot encoding format and $\tilde{y}$ is the label corresponding to the mixed batch sample. The processed image is shown in Fig. 7.7.

(4) Cutout Augmentation

It is unavoidable that some targets are covered in the practical ground fissure detecting process. To avoid the situation that the ground fissures are overlooked by the detecting system, the model is required to have ability to detect the covered targets. In this case, the original samples can be augmented by Cutout. The principle of Cutout

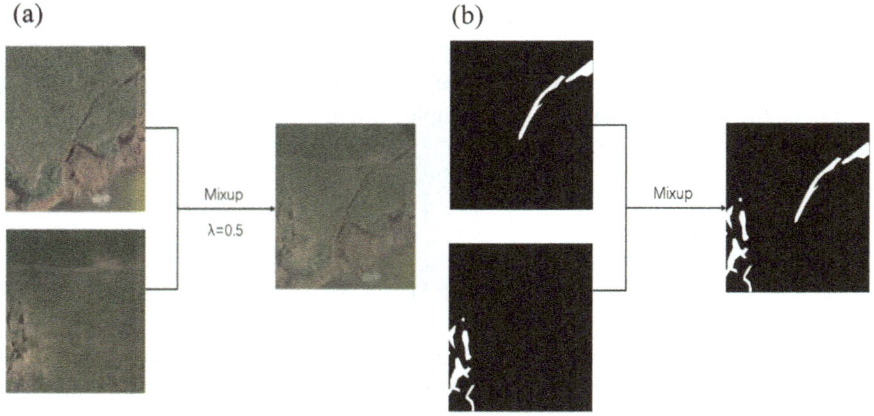


**Fig. 7.7** Image mixup augmentation (**a**) and label mixup augmentation (**b**)

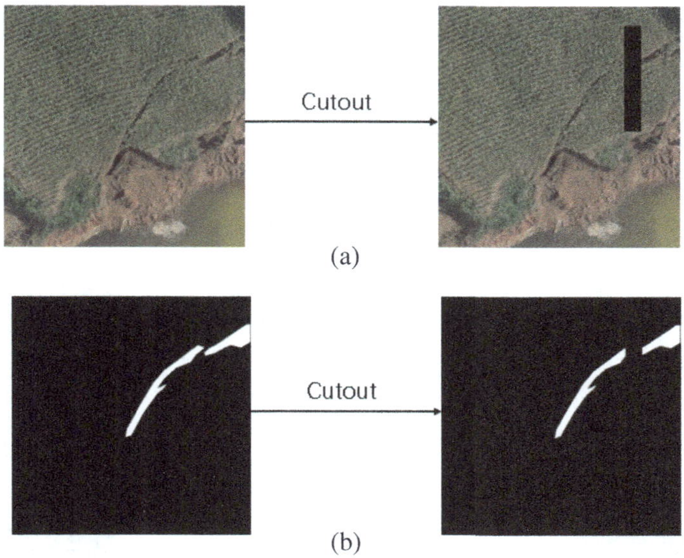


**Fig. 7.8** Cutout augmentation of **a** image and **b** label

augmentation (DeVries & Taylor, 2017) is to randomly select a square area with fixed size and fill it with zeros. The image processed by Cutout is shown in Fig. 7.8.

(5) Mosaic Augmentation

Mosaic augmentation is proposed in Yolo v4 (Bochkovskiy et al., 2020). The principle is to randomly choose four images of each batch from the training dataset and carry out operations including rotation, scaling, and cropping, and then integrate them into one image by following the order of upper left, upper right, lower left and lower right. Mosaic augmentation can balance the distribution of objects in different scales during the training process and meanwhile enrich the background of the detected objects, which is conducive to improving the detecting performance of small targets. In addition, the splicing of the four sub-images can increase the batch size of each input batch, which reduces the dependence on batch size, as well as the requirement for video memory. Image processed by Mosaic augmentation is shown in Fig. 7.9.

### *7.3.3 Dataset Construction*

After the abovementioned operations are carried out on the acquired data, 15,000 images with a resolution of 256 × 256 are finally obtained to make the ground fissure dataset. The constructed dataset includes three different vegetation coverage situations: "covered with vegetation", "partially covered with vegetation", and "no vegetation cover". Here, the dataset is divided into three parts for training, testing

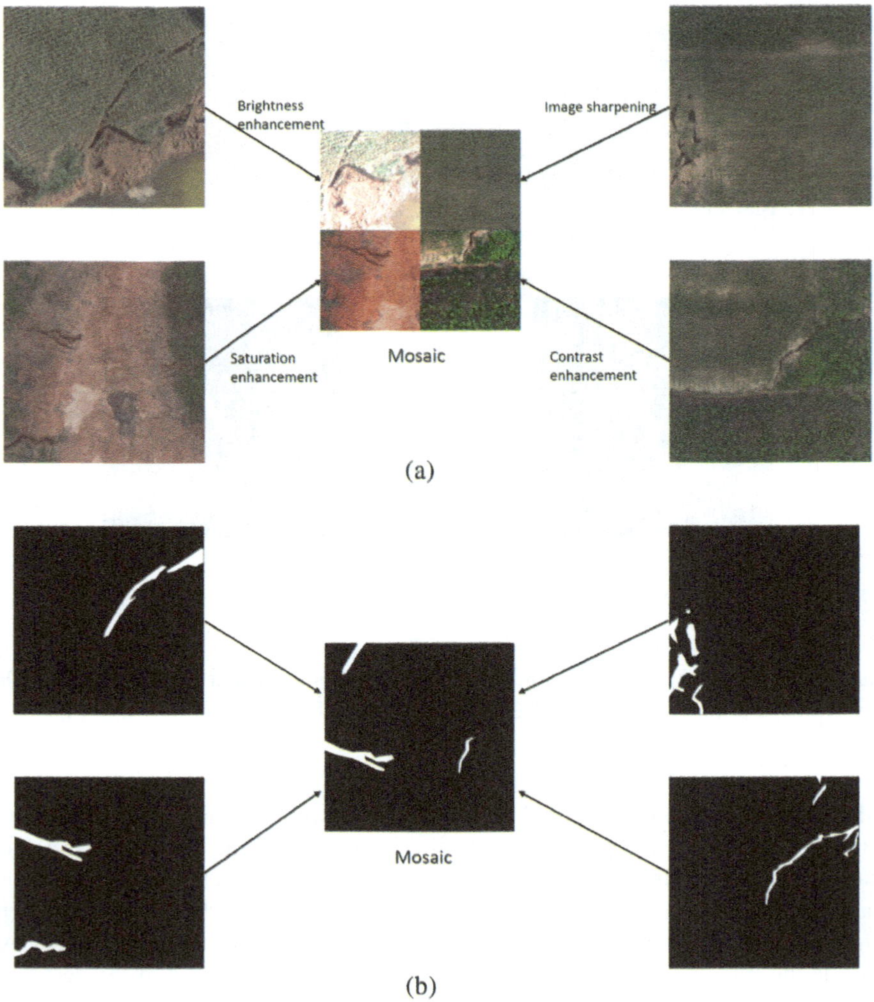


**Fig. 7.9** Mosaic augmentation of **a** image and **b** label

and validation. The training dataset contains 10,000 images, the validation dataset contains 2000 images, and the test dataset contains 3000 images. The image example of the ground fissure dataset is shown in Fig. 7.10.

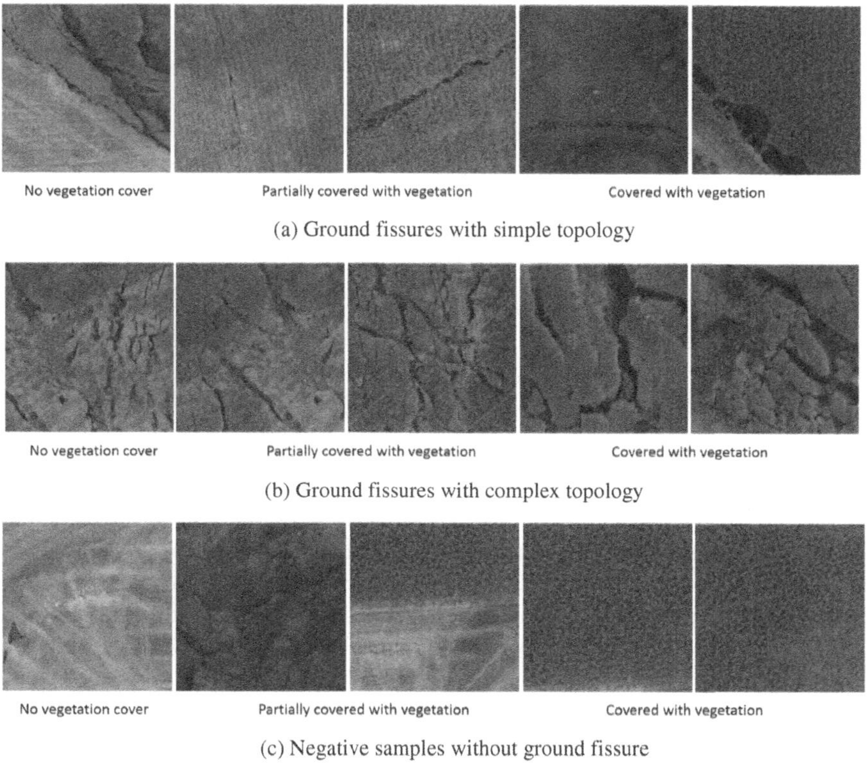


**Fig. 7.10** Ground fissure dataset

## 7.4 Ground Fissure Segmentation Based on Multiscale Graph Convolution Feature

### *7.4.1 Segmentation Framework*

The segmentation framework proposed in this research is shown in Fig. 7.11. The whole framework is trained in an end-to-end manner. In the basis of U-Net segmentation network, a multi-scale global reasoning module (Li et al., 2021) is included. Therefore, the network is composed of encoder, multi-scale global reasoning module and decoder.

The first part is the encoder for feature extraction. Two convolution layers with kernel size of 3 × 3 perform the feature extraction from the input image. Each convolution layer is followed by a batch normalization (BN) layer and a ReLU activation function. Then the max pooling layer with 2 × 2 kernel size is used to decrease the size of the feature map. The encoder performs the abovementioned operations for three times.

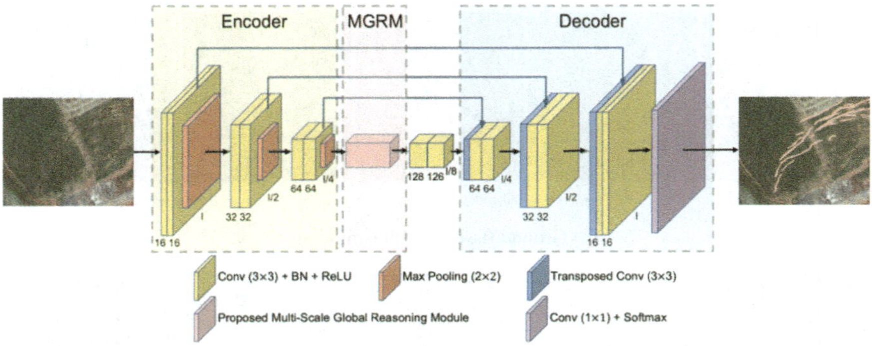


**Fig. 7.11** Proposed framework for ground fissure segmentation

The second part is a multi-scale global reasoning module. The detailed structure and function will be introduced in Sect. 7.4.2.

The third part is the decoder for image upsampling. The transpose convolution with kernel size of 3 × 3 is used to generate a feature map that is of twice width and height of the encoded one. Then, the feature map obtained from corresponding layers in the feature extraction stage is cropped, copied, and spliced with the feature map obtained by upsampling. Two convolution layers with kernel size of 3 × 3 then perform the feature extraction of spliced feature map. Each convolution layer is followed by a BN layer and a ReLU activation function. The decoder also performs the above operations for three times.

Finally, the softmax activation function is used to transform the feature vectors to probability distribution vectors, and output the final segmented image.

### Batch Normalization

The learning of data distribution is the essence of neural network learning. In the training process, the variation of parameters can significantly affect the following data. The tiny changes of data in the front layer can be amplified with the propagation process, resulting in very different output results. The high training complexity is caused by the instability of numerical calculation, which can lead to lower training efficiency and higher risk of overfitting.

To tackle this issue, a layer for the normalization of the output is considered before the activation function, so that the data can be in the same distribution before moving to the next layer. This method can improve the learning and training efficiency of neural network and reduce the overfitting risk. The batch normalization (Ioffe & Szegedy, 2015) can be expressed from Eqs. (7.13) to (7.16):

$$\mu_{\mathrm{B}} = \frac{1}{m}\sum_{i=1}^{m} x_i \tag{7.13}$$

$$\sigma_{\mathrm{B}}^2 = \frac{1}{m}\sum_{i=1}^{m}(x_i - \mu_{\mathrm{B}})^2 \tag{7.14}$$

$$\widehat{x_i} = \frac{x_i - \mu_B}{\sqrt{\sigma_B^2 + \epsilon}} \tag{7.15}$$

$$y_i = \gamma \widehat{x_i} + \beta \equiv BN_{\gamma,\beta}(x_i) \tag{7.16}$$

where $m$ represents the data size of a batch. $\mu_B$ is the mean value of data in a batch. $\sigma_B^2$ represents the variance of data in a batch. $\widehat{x_i}$ represents the normalized data value. $\epsilon$ is a smoothing factor that avoids denominators being zero, $\gamma$, $\beta$ are parameters of the network.

**Transpose Convolution**

Transposed convolution (Long et al., 2015) is also called deconvolution, which is the inverse process of convolution in deep learning. The size of the input image tends to become smaller when it is output because of the feature extraction process of CNN. The function of transpose convolution is to restore the size of the image by setting convolution kernel size and the output size. UpSampling2D function is the sampling function used in traditional U-Net. However, this function does not contain trainable parameters, and only performs data interpolation and multiplication on image data in the directions of width and height. In this study, UpSampling2D is replaced by transposed convolution Con2DTranspose, which has learning ability in the image restore stage, to improve the model ability to output the detailed features. The process of transpose convolution is shown in Fig. 7.12.

**Latent Space**

Latent space (White, 2016) is a representation of compressed data. The function of latent space is to find patterns by learning data features and simplify data representation. The purpose of data compression is to learn more important information in data. The overall framework of this study is an encoder-decoder network. First, the full CNN is used to learn image features, and the dimensionality reduction of data in feature extraction can be regarded as a lossy compression of data. In the process of data reconstruction by the decoder, the model needs to learn how to store as much as relevant information and meanwhile ignore noise. Therefore, the advantage of data

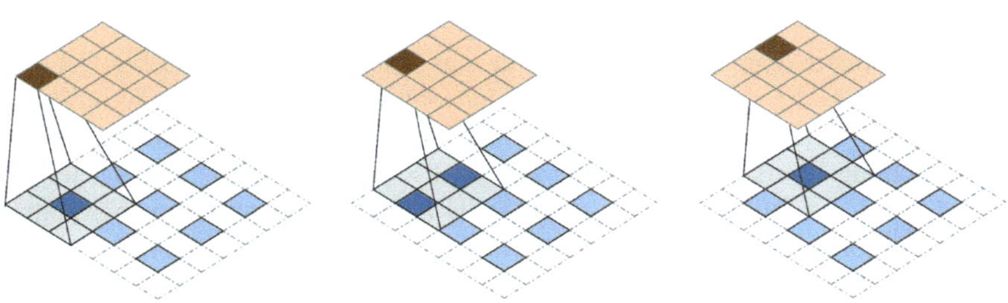

**Fig. 7.12** Process of transpose convolution

compression is that it can remove redundant information and focus on the features of key data. This compressed state is the latent space representation of data.

### 7.4.2 Multiscale Global Reasoning Module

Based on the U-net, this study proposes a multi-scale graph convolution reasoning module (MGRB). MGRB is in the center of the network and connects the encoder and decoder. It is used to obtain a large scale of context features that are difficult to be obtained by traditional neural networks to conduct global reasoning on the high-level semantic features of ground fissures.

MGRB is composed of multi-scale pooling operator and graph reasoning module. Multiple effective receptive fields are used to learn the features of ground fissures in different sizes. Receptive field (Luo et al., 2016) is defined as the size of the area on the input image mapped back by the pixels on the feature map output from each layer in CNN. Therefore, the pixels on feature map are calculated from the receptive field in the input image. The equations for calculating the receptive field of a specific layer are expressed as Eqs. (7.17) and (7.18).

$$RF_{i+1} = RF_i + (k - 1) \times S_i \tag{7.17}$$

$$S_i = \prod_{i=1}^{i} Stride_i \tag{7.18}$$

where, the size of receptive field of the specific layer is represented by $RF_{i+1}$, and $RF_i$ represents the size of the receptive field of the last layer. $k$ represents the size of the convolution kernel. $S_i$ is the product of strides of all previous layers. The range of original image can be reached by neurons is positively correlated with the value of receptive field. The higher value of receptive field means the better global features can be obtained. In contrast, there is a negative correlation between the image local details and the value of receptive field. The lower value of receptive field allows the image restoration to be more detailed. Workflow of receptive field is shown in Fig. 7.13.

The schematic diagram of MGRB with four branches is shown in Fig. 7.14. In the first branch, the data processed by a convolution with a kernel size of 3 × 3 is directly delivered to the graphic reasoning module. Based on Ce-Net (Gu et al., 2019), pooling layers with the kernel size of 2 × 2, 3 × 3, 5 × 5 are set for the second, third and fourth branches respectively before the graph reasoning module (The details of graph reasoning module will be introduced in Sect. 7.4.3). Then, to fit the size of the input feature map, the global features are up sampled by bilinear interpolation. Finally, the features from four branches are recombined and connected to obtain the final output feature map.

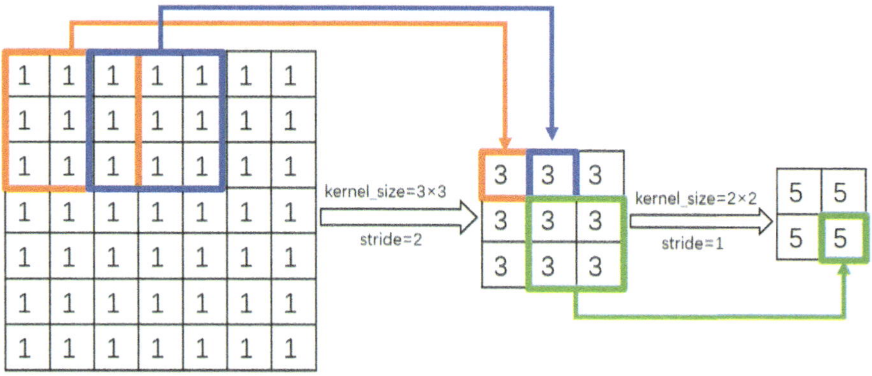


**Fig. 7.13** Schematic diagram of receptive field

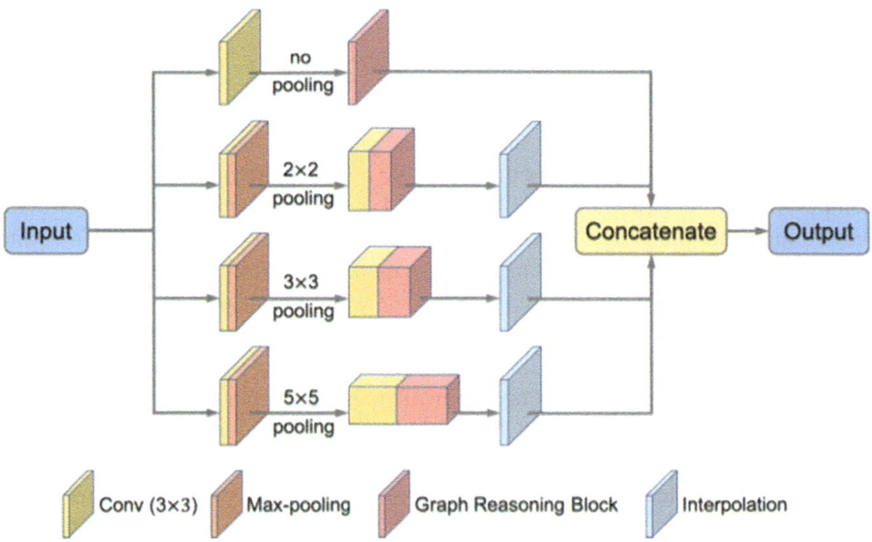


**Fig. 7.14** Schematic diagram of multi-scale global reasoning module

## *7.4.3 Graph Reasoning Module*

In the MGRB module of GCN, the key part is the graph reasoning module. Based on previous researches of graph-based global reasoning networks (Chen et al., 2019) and spatial pyramid based graph reasoning for semantic segmentation (Li et al., 2020), a graph reasoning module is designed in this study. The architecture of graph reasoning module is shown in Fig. 7.15, which can effectively extract the global features of ground fissure.

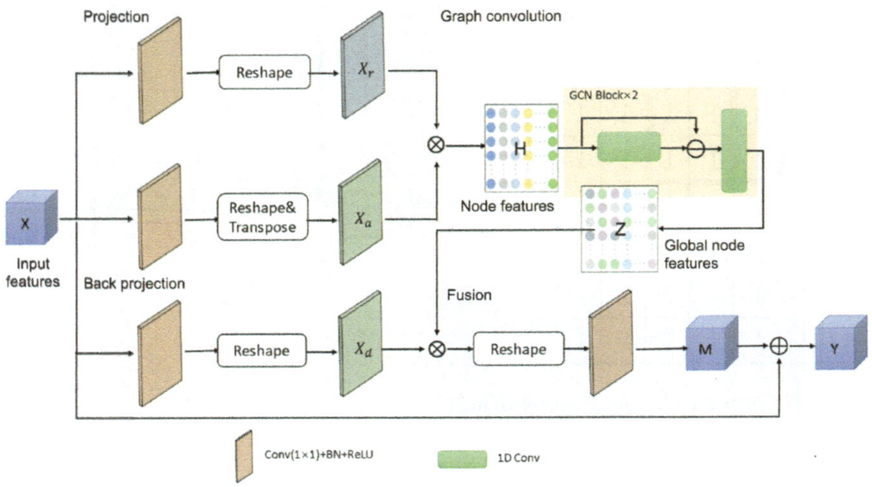


**Fig. 7.15** Schematic diagram of graph reasoning module

The graph reasoning module includes four parts. The first part is to project the original feature into the node space. Firstly, the local feature map in the latent space is input into two convolution layers with 1 × 1 kernels in parallel. Afterwards, two feature maps are generated: one is the feature map after dimension reduction, and the other one is the projection matrix feature map. The feature map after dimension reduction is reconstructed into $X_r \in R^{C_r \times H \times W}$, and the projection matrix feature map is reconstructed and transposed into $X_a \in R^{C_n \times H \times W}$. Matrix multiplication is performed for $X_r$ and $X_a$ to obtain a node feature mapping matrix $H \in R^{C_r \times C_n}$.

The second part is the graph convolution of the features of the node space to extract the features of the global nodes. GCN is one of the best modules for processing graph data. It uses message propagation to exchange information between adjacent nodes, which can extract features in a large adjacent range of the graph. Since no nodes will disappear in this process, GCN based on nodes can expand the receptive field and avoid the loss of local location information. Graph data is usually in triple form $G(N, E, U)$. $N$ represents the set of graph nodes. $E$ is the set of graph edges, which is represented by the adjacency matrix. $U$ is the feature of graph. The feature of graph is not used in the task, so it is omitted. A two-layer GCN model is constructed in this study. All of the activation functions are ReLU. The forward propagation formula of GCN is shown in Eq. (7.19):

$$Z = f(X, A) = ReLU(\hat{A}ReLU\left(\hat{A}XW^{(0)}\right)W^{(1)} \quad (7.19)$$

$$\hat{A} = A + I \quad (7.20)$$

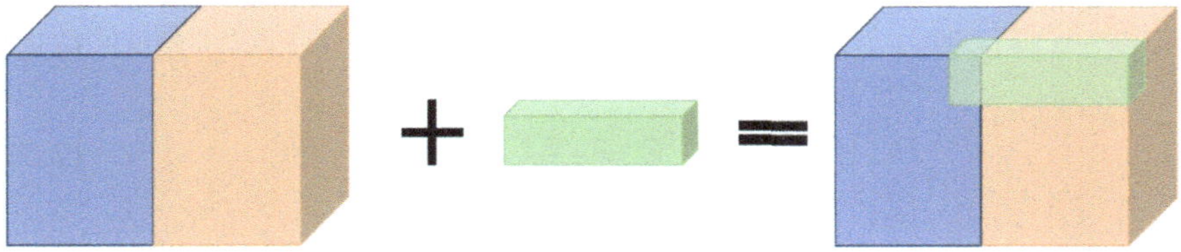

**Fig. 7.16** Workflow of element-wise addition feature fusion

where $X$ is the input node feature, which is the node feature mapping matrix $H$ obtained in the first part. $\widehat{A}$ is the sum of adjacency matrix and identity matrix. $W^{(0)}$ is the parameter in first layer of GCN, and $W^{(1)}$ is the parameter in second layer of GCN. Finally, the eigenmatrix of the second-order adjacent information is obtained.

In the third part, in order to fuse the global node features with the original features, the global node features are back projected into the original feature space. The original input feature matrix $X$ is connected to the convolution layer with $1 \times 1$ kernel to create an inverse projection matrix $X_d \in R^{C_n \times H \times W}$.

The fourth part is the fusion of global node features and original features. To transform back to the original latent space, the global node feature matrix obtained from the GCN is multiplied by matrix $X_d$ obtained in third part. The product is processed by $1 \times 1$ convolution layer to obtain the feature map matrix $M \in R^{C \times H \times W}$. Finally, the features of matrix M and matrix X are fused by element-wise addition operation. feature maps are added without changing the number of channels to obtain a new feature map $Y \in R^{C \times H \times W}$. The feature fusion process is shown in Fig. 7.16.

In this study, the classical binary cross entropy function is used as the loss function of the ground fracture detection network, and it can be calculated by Eq. (7.21):

$$L_{BCE} = -\frac{1}{N}\sum_{i=1}^{N}\left[y_i \log p_i + (1 - y_i)\log(1 - p_i)\right] \tag{7.21}$$

where $N$ represents the number of pixels in the image. $y_i$ represents the labeled value of the pixel $i$. $p_i$ represents the prediction probability of the pixel $i$.

Adam algorithm (Kingma & Ba, 2014) can be used in the optimization task with large amount of data and parameters with lower requirement of memory. This research adopts Adam method to adjust model parameters. The update process of Adam optimization algorithm is shown from Eqs. (7.22) to (7.27):

$$g_t = \nabla_\theta f_t(\theta_t - 1) \tag{7.22}$$

$$m_t = \beta_1 \cdot m_{t-1} + (1 - \beta_1) \cdot g_t \tag{7.23}$$

$$v_t = \beta_2 \cdot v_{t-1} + (1 - \beta_2) \cdot g_t^2 \tag{7.24}$$

$$\widehat{m_t} = \frac{m_t}{1-\beta_1^t} \tag{7.25}$$

$$\widehat{v_t} = \frac{v_t}{1-\beta_2^t} \tag{7.26}$$

$$\theta_t = \theta_{t-1} - \alpha \cdot \frac{\widehat{m_t}}{\sqrt{\widehat{v_t}+\epsilon}} \tag{7.27}$$

where $t$ represents the timestep. $\alpha$ represents the learning rate, which is used to control the stepsize. It is set to 0.001 in this study. $\theta$ represents the parameter to be updated. $f(\theta)$ represents an objective function with the parameter $\theta$. $g_t$ denotes the gradient obtained by the derivative of $f(\theta)$ to $\theta$. $\beta_1$ represents the first-order moment attenuation coefficient, which is set to 0.9. $\beta_2$ represents the second-order moment attenuation coefficient, which is set to 0.99. $m_t$ represents the first moment of the gradient $g_t$. $v_t$ represents the second moment of the gradient $g_t$. $\widehat{m_t}$ represents the bias-corrected first moment and $\widehat{v_t}$ represents the bias-corrected second moment.

## 7.5 Experimental Results and Analysis

### *7.5.1 Experimental Environment*

All experiments are carried out under the unified computational environment to ensure the accuracy and effectiveness of test data. The details of hardware and software of relevant computers are shown in Table 7.1.

**Table 7.1** Configuration of experimental environment

| Experimental requirements | Configuration | Parameter |
|---|---|---|
| Hardware | Operating system | Windows 10 bit-64 |
| | Platform | Python 3.9 |
| | Compiler | PyCharm |
| | CPU | Intel(R)Core(TM)i5-8300H CPU@2.3 GHz |
| | GPU | NVIDIA GeForce GTX 1050 Ti |
| Software | Pytorch | 1.11.0 |
| | CUDA | 11.1 |
| | cuDNN | 8.0.0 |

### 7.5.2 Results and Analysis

#### Evaluation Metrics

This study evaluates the segmentation performance of the model by comparing the segmentation results of the model with the of manual annotation. S is used to represent the pixel set of the target area, which is obtained by a series of segmentation algorithms. The pixel set of manually labeled target area is represented by P. Then the precision, recall and f1-score are selected for evaluation. The calculation method is as follows:

Precision indicates the proportion of true positive category in all samples determined as positive category. It can be calculated by Eq. (7.28):

$$Precision = \frac{S \cap P}{S} \tag{7.28}$$

Recall represents the proportion of all the real positive samples determined as positive category. It can be calculated by Eq. (7.29):

$$Recall = \frac{S \cap P}{P} \tag{7.29}$$

F1-score represents the harmonic average of precision coefficient and recall coefficient and can be calculated by Eq. (7.30):

$$F1\ Score = 2 \times \frac{Precision \times Recall}{Precision + Recall} \tag{7.30}$$

#### Comparative Experiments

In this research, firstly, the traditional digital image processing method is used to extract the ground fissures in the experimental dataset to make comparison of the method performances, including Canny edge detection and region growth algorithm edge detection. In addition, the DL-based semantic segmentation models are also compared with the proposed model, including DeepLab v3, PSPNet (Pyramid Scene Parsing Network), and U-Net. The network structure is described as follows.

(1) Canny edge detection algorithm (Canny, 1986): Gaussian filter is used to smooth the image and remove the noise. Afterwards, Sobel operator is used to calculate the gradient intensity and direction of the pixel. The spurious response caused by edge detection is suppressed and eliminated by non-maximum pixels. And the detected possible edge is determined by double threshold method. The final image edge is tracked by hysteresis technology.
(2) Region growth algorithm (Chen, 2008): All pixels in the image are divided into attributed pixels and non-attributed pixels. The first non-attributed pixel is found by the method of image sequential scanning. Taking this pixel as the central pixel, the adjacent pixels that meet the growth criteria are merged into a region,

and all pixels in this region are marked as attributed pixels. The abovementioned steps are repeated until all pixels in the image are changed to attributed pixels.

(3) U-Net: The introduction of U-Net can be found in Sect. 7.2.1.
(4) PSPNet (Zhao et al., 2017): A pyramid pooling module based on integration of global context information is proposed. This global prior information can effectively obtain high-quality results in scene semantic analysis. Effective optimization strategy is developed for deep ResNet based on the loss of deep supervised learning.
(5) DeepLab v3 (Chen et al., 2018): The problem of multi-scale segmentation of objects is solved by cascading or parallel methods. The hole convolution with different hole rates can capture the context information in different scales. In addition, DeepLab v3 extends the hole convolution space pyramid pooling module. The model can not only encode the global context features of the image dataset, but also improve the performance of the model by detecting convolution features in multiple scales.

The comparison of ground fissure segmentation results in different models is shown in Fig. 7.17, and the variations of loss in the training process of four deep learning models are shown in Fig. 7.18. In the experiment, the original images containing ground fissures in Fig. 7.17a include the situations with complex background, shadow, noise, small fissures and complex topology, which are all factors affecting the performance of ground fissure detection. According to the performances of comparative experiments in Fig. 7.17b–g, it can be concluded that the traditional methods of Canny edge detection and region growth edge detection algorithms are not suitable for processing the images in this study, with a large amount of extracted edge noise and low performance. Regarding the DL-based methods, U-Net performed much better than the traditional methods, but still failed to extract complete ground fissures. PSPNet is not sensitive to noise interference, but due to the multi-scale feature optimization is not included, micro ground fissures cannot be always successfully detected. DeepLab v3 model can detect the micro ground fissures due to its hole convolution with different void rates to obtain multi-scale information of the image. However, some shadows of vegetation are also recognized as ground fissures. The model proposed in this chapter integrates convolutions in different scales to obtain low-level and high-level features. Inclusion of GCN network learning allows the model to have higher ground fissure detection accuracy and reduces the influences of noise.

Table 7.2 indicates the performance comparison of ground fissure segmentation in different deep learning models and the highest values of evaluation metrics for segmenting images with fissure are bolded. The proposed method indicates the highest performance in precision and F1-score, reaching 0.76 and 0.65 respectively. And the recall is also very close to the highest value gained in PSPNet. Overall, it is the most suitable model to process the ground fissure dataset.

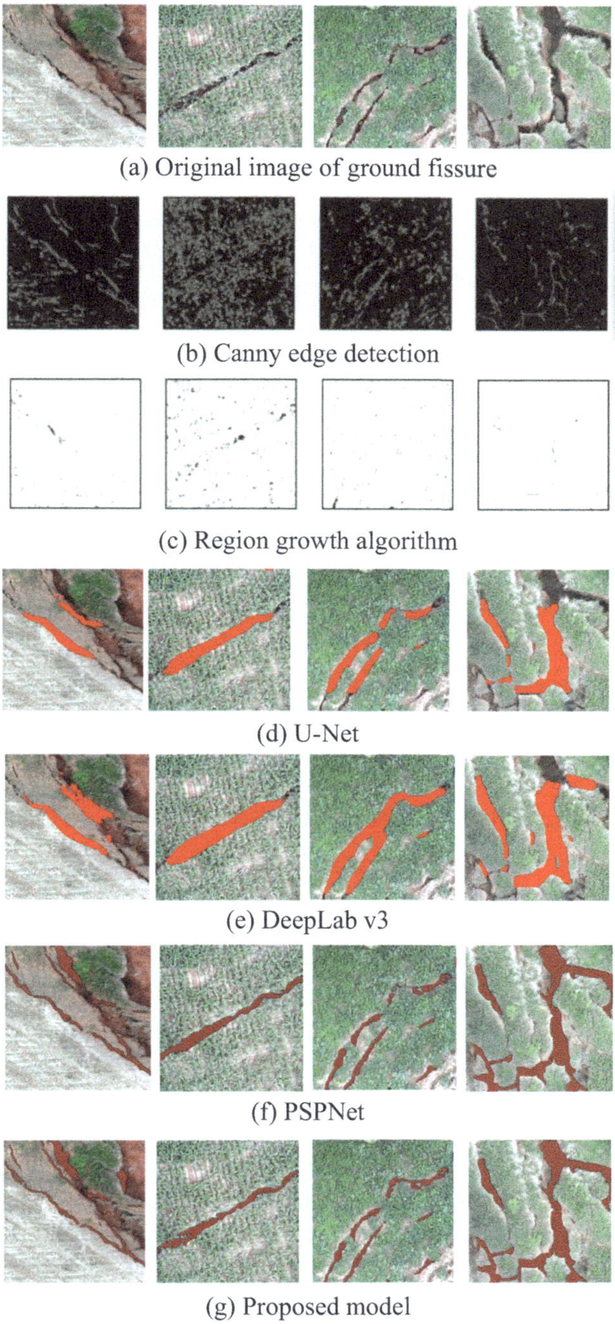


**Fig. 7.17** Performance comparison of different ground fissure segmentation methods

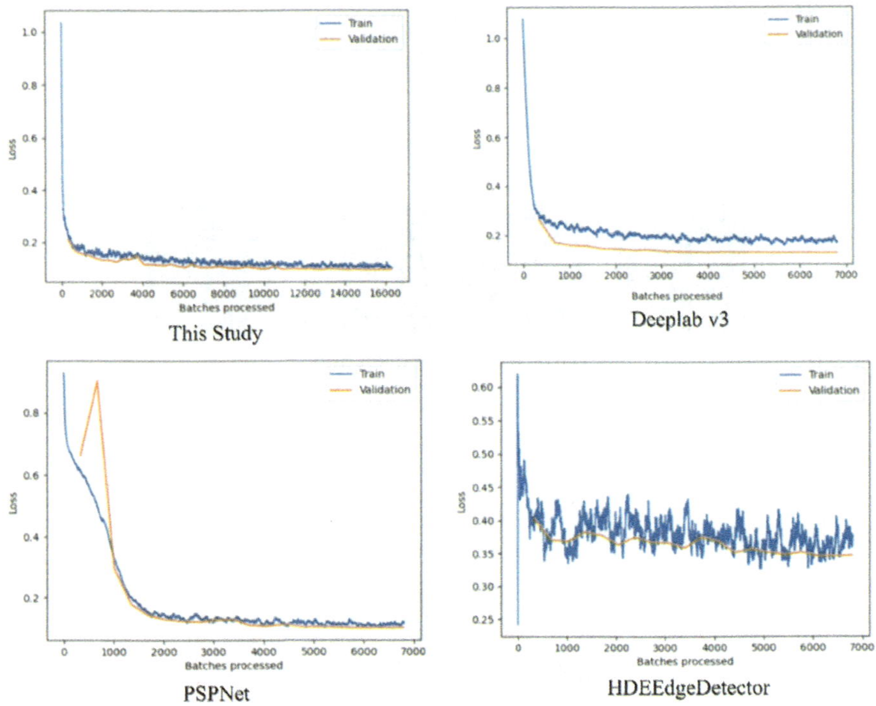


**Fig. 7.18** Variation of loss in training processes of deep learning models

**Table 7.2** Comparison results of segmentation effect evaluation coefficients of different models

| Methods | Precision | | Recall | | F1-score | |
|---|---|---|---|---|---|---|
| | No fissure | Fissure | No fissure | Fissure | No fissure | Fissure |
| U-net | 0.94 | 0.59 | 0.92 | 0.35 | 0.93 | 0.44 |
| DeepLabv3 | 0.96 | 0.59 | 0.98 | 0.35 | 0.97 | 0.44 |
| PSPNet | 0.97 | 0.72 | 0.99 | **0.57** | 0.98 | 0.64 |
| Paper model | 0.97 | **0.76** | 0.99 | 0.57 | 0.98 | **0.65** |

## 7.6 Summary

This chapter proposed a U-shaped model for automatic detection of ground fissures with complex backgrounds based on multi-scale graph convolution feature. The remote sensing images of regions in Shanxi Province are used to construct the ground fissures dataset. The dataset contains samples with different vegetation coverage, which can represent the most situations of ground fissures.

Using the multiscale convolution technology to obtain and fuse the high-level and low-level information in the ground fissure image can effectively identify the micro ground fissures. In order to detect ground fissures with complex topologic structure, which usually does not have translation invariance and makes the traditional CNN and RNN invalid, GCN is introduced to learn the graph data with irregular structure. Finally, the performance of the proposed model is tested and proved to be feasible and effective in the ground fissure segmentation task.

The DL-based ground fissure segmentation model prosed in this study can be improved from following aspects:

(1) Attention mechanism can be introduced into the network to learn more targeted and discriminative ground fissure features.
(2) GCN model is used for fusion in this study. Subsequent researches can consider other graph network structures to explore a network structure more suitable for ground fissure extraction.
(3) The proposed model cannot achieve the real-time processing of the collected remote sensing images. The automatic real-time monitoring method of ground fissures can be studied to improve the detection efficiency.

### Codes

### [Python3.9] MGUNet

```
import torch
import torch.nn as nn
import torch.nn.functional as F
from models.utils.utils import Basconv, UnetConv, UnetUp, UnetUp4,
GloRe_Unit
from collections import OrderedDict
from models.utils.init_weights import init_weights
class MGR_Module(nn.Module):
    def __init__(self, in_channels, out_channels):
        super(MGR_Module, self).__init__()
            self.conv0_1 = Basconv(in_channels=in_channels, out_
channels=out_channels, kernel_size=3, padding=1)
              self.glou0 = nn.Sequential(OrderedDict([("GCN%02d"
% i, GloRe_Unit(out_channels, out_channels, kernel=1)) for i in
range(1)]))
             self.conv1_1 = Basconv(in_channels=in_channels,out_
channels=out_channels, kernel_size=3, padding=1)
        self.pool1 = nn.MaxPool2d(kernel_size=[2, 2], stride=2)
            self.conv1_2 = Basconv(in_channels=out_channels, out_
channels=out_channels, kernel_size=3, padding=1)
              self.glou1 = nn.Sequential(OrderedDict([("GCN%02d"
% i,GloRe_Unit(out_channels, out_channels, kernel=1)) for i in
range(1)]))
            self.conv2_1 = Basconv(in_channels=in_channels, out_
channels=out_channels, kernel_size=3, padding=1)
        self.pool2 = nn.MaxPool2d(kernel_size=[3, 3], stride=3)
            self.conv2_2 = Basconv(in_channels=out_channels, out_
channels=out_channels, kernel_size=3, padding=1)
```

```
            self.glou2 = nn.Sequential(OrderedDict([("GCN%02d" %
i,GloRe_Unit(out_channels, int(out_channels/2), kernel=1)) for i
in range(1)]))
            self.conv3_1 = Basconv(in_channels=in_channels, out_
channels=out_channels, kernel_size=3, padding=1)
        self.pool3 = nn.MaxPool2d(kernel_size=[5, 5], stride=5)
           self.conv3_2 = Basconv(in_channels=out_channels, out_
channels=out_channels, kernel_size=3, padding=1)
            self.glou3 = nn.Sequential(OrderedDict([("GCN%02d" %
i,GloRe_Unit(out_channels, int(out_channels/2), kernel=1)) for i
in range(1)]))
             self.f1 = Basconv(in_channels=4*out_channels, out_
channels=in_channels, kernel_size=1, padding=0)
    def forward(self, x):
        self.in_channels, h, w = x.size(1), x.size(2), x.size(3)
        self.x0 = self.conv0_1(x)
        self.g0 = self.glou0(self.x0)
        self.x1 = self.conv1_2(self.pool1(self.conv1_1(x)))
        self.g1 = self.glou1(self.x1)
             self.layer1 = F.interpolate(self.g1, size=(h, w),
mode='bilinear', align_corners=True)
        self.x2 = self.conv2_2(self.pool2(self.conv2_1(x)))
        self.g2 = self.glou2(self.x2)
             self.layer2 = F.interpolate(self.g2, size=(h, w),
mode='bilinear', align_corners=True)
        self.x3 = self.conv3_2(self.pool3(self.conv3_1(x)))
        self.g3= self.glou3(self.x3)
             self.layer3 = F.interpolate(self.g3, size=(h, w),
mode='bilinear', align_corners=True)
            out = torch.cat([self.g0, self.layer1, self.layer2,
self.layer3], 1)
        return self.f1(out)
class MGUNet_1(nn.Module):
      def __init__(self, in_channels=1, n_classes=11, feature_
scale=4, is_deconv=True, is_batchnorm=True): ##########
        super(MGUNet_1, self).__init__()
        self.is_deconv = is_deconv
        self.in_channels = in_channels
        self.is_batchnorm = is_batchnorm
        self.feature_scale = feature_scale
        filters = [64, 128, 256, 512, 1024]
        filters = [int(x / self.feature_scale) for x in filters]
        # encoder
      self.conv1 = UnetConv(self.in_channels, filters[0], self.is_
batchnorm)
        self.maxpool1 = nn.MaxPool2d(kernel_size=2)
          self.conv2 = UnetConv(filters[0], filters[1], self.is_
batchnorm)
        self.maxpool2 = nn.MaxPool2d(kernel_size=4)
          self.conv3 = UnetConv(filters[1], filters[2], self.is_
batchnorm)
        self.maxpool3 = nn.MaxPool2d(kernel_size=4)
        self.mgb = MGR_Module(filters[2], filters[3])
```

```
        self.center = UnetConv(filters[2], filters[3], self.is_
batchnorm)
        # decoder
        self.up_concat3 = UnetUp4(filters[3], filters[2], self.is_
deconv)
        self.up_concat2 = UnetUp4(filters[2], filters[1], self.is_
deconv)
        self.up_concat1 = UnetUp(filters[1], filters[0], self.is_
deconv)
        # final conv
        self.final_1 = nn.Conv2d(filters[0], n_classes, 1)
        # initialise weights
        for m in self.modules():
            if isinstance(m, nn.Conv2d):
                init_weights(m, init_type='kaiming')
            elif isinstance(m, nn.BatchNorm2d):
                init_weights(m, init_type='kaiming')
    def forward(self, inputs):
        conv1 = self.conv1(inputs)
        maxpool1 = self.maxpool1(conv1)
        conv2 = self.conv2(maxpool1)
        maxpool2 = self.maxpool2(conv2)
        conv3 = self.conv3(maxpool2)
        maxpool3 = self.maxpool3(conv3)
        feat_sum = self.mgb(maxpool3)
        center = self.center(feat_sum)
        up3 = self.up_concat3(center, conv3)
        up2 = self.up_concat2(up3, conv2)
        up1 = self.up_concat1(up2, conv1)
        final_1 = self.final_1(up1)
        return final_1
class MGUNet_2(nn.Module):
    def __init__(self, in_channels=1, n_classes=11, feature_
scale=4, is_deconv=True, is_batchnorm=True): ##########
        super(MGUNet_2, self).__init__()
        self.is_deconv = is_deconv
        self.in_channels = in_channels
        self.is_batchnorm = is_batchnorm
        self.feature_scale = feature_scale
        filters = [64, 128, 256, 512, 1024]
        filters = [int(x / self.feature_scale) for x in filters]
        # encoder
        self.conv1 = UnetConv(self.in_channels, filters[0], self.is_
batchnorm)
        self.maxpool1 = nn.MaxPool2d(kernel_size=2)
        self.conv2 = UnetConv(filters[0], filters[1], self.is_
batchnorm)
        self.maxpool2 = nn.MaxPool2d(kernel_size=2)
        self.conv3 = UnetConv(filters[1], filters[2], self.is_
batchnorm)
        self.maxpool3 = nn.MaxPool2d(kernel_size=2)
        self.mgb = MGR_Module(filters[2], filters[3])
        self.center = UnetConv(filters[2], filters[3], self.is_
batchnorm)
```

```
        # decoder
        self.up_concat3 = UnetUp(filters[3], filters[2], self.is_
deconv)
        self.up_concat2 = UnetUp(filters[2], filters[1], self.is_
deconv)
        self.up_concat1 = UnetUp(filters[1], filters[0], self.is_
deconv)
        # final conv
        self.final_1 = nn.Conv2d(filters[0], n_classes, 1)
        # initialise weights
        for m in self.modules():
            if isinstance(m, nn.Conv2d):
                init_weights(m, init_type='kaiming')
            elif isinstance(m, nn.BatchNorm2d):
                init_weights(m, init_type='kaiming')
    def forward(self, inputs):
        conv1 = self.conv1(inputs)
        maxpool1 = self.maxpool1(conv1)
        conv2 = self.conv2(maxpool1)
        maxpool2 = self.maxpool2(conv2)
        conv3 = self.conv3(maxpool2)
        maxpool3 = self.maxpool3(conv3)
        feat_sum = self.mgb(maxpool3)
        center = self.center(feat_sum)
        up3 = self.up_concat3(center, conv3)
        up2 = self.up_concat2(up3, conv2)
        up1 = self.up_concat1(up2, conv1)
        final_1 = self.final_1(up1)
        return final_1
```

### [Python3.9] OSNet

```
import torch
import torch.nn as nn
import copy
import torch.nn.functional as F
from models.nets.MGUNet import MGUNet_2
class OSMGUNet(nn.Module):
    def __init__(self): ##########
        super(OSMGUNet, self).__init__()
       self.stage = MGUNet_2(in_channels=1, n_classes=11, feature_
scale=4)
    def forward(self, inputs):
        out = self.stage(inputs)
        output = F.log_softmax(out, dim=1)
        return output
```

### [Python3.9] TSNet

```
import torch
import torch.nn as nn
import copy
```

```
import torch.nn.functional as F
from models.nets.MGUNet import MGUNet_1, MGUNet_2
from models.utils.utils import img2df, feature_fusion
class TSMGUNet(nn.Module):
    def __init__(self):
        super(TSMGUNet, self).__init__()
       self.stage1 = MGUNet_1(in_channels=1, n_classes=3, feature_
scale=4)
      self.stage2 = MGUNet_2(in_channels=1, n_classes=10, feature_
scale=4)
    def forward(self, inputs):
        input_layer2 = copy.deepcopy(inputs)
        out1 = self.stage1(inputs)
        output1 = F.log_softmax(out1, dim=1)
        _, pred = torch.max(output1, 1)
        input_mask = (pred).unsqueeze(1).float()
        input_disc_free = img2df(input_layer2,input_mask)
        out2 = self.stage2(input_disc_free)
        output2, out = feature_fusion(out1, out2)
        output = F.log_softmax(out, dim=1)
        return output1, output2, output
```

**[Python3.9] Model_utils**

```
import torch
import torch.nn as nn
import torch.nn.functional as F
from init_weights import init_weights
class Basconv(nn.Sequential):
     def __init__(self, in_channels, out_channels, is_batchnorm =
False, kernel_size = 3, stride = 1, padding=1):
        super(Basconv, self).__init__()
        if is_batchnorm:
                  self.conv = nn.Sequential(nn.Conv2d(in_channels,
out_channels, kernel_size, stride, padding),nn.BatchNorm2d(out_
channels),nn.ReLU(inplace=True))
        else:
            self.conv = nn.Sequential(nn.Conv2d(in_channels, out_
channels, kernel_size, stride, padding),nn.ReLU(inplace=True))
        # initialise the blocks
        for m in self.children():
            init_weights(m, init_type='kaiming')
    def forward(self, inputs):
        x = inputs
        x = self.conv(x)
        return x
class UnetConv(nn.Module):
      def __init__(self, in_channels, out_channels, is_batchnorm,
n=2, kernel_size = 3, stride=1, padding=1):
        super(UnetConv, self).__init__()
        self.n = n
        if is_batchnorm:
            for i in range(1, n+1):
```

```
                conv = nn.Sequential(nn.Conv2d(in_channels, out_
channels, kernel_size, stride, padding),
                                     nn.BatchNorm2d(out_channels),
                                     nn.ReLU(inplace=True),)
                setattr(self, 'conv%d'%i, conv)
                in_channels = out_channels
        else:
            for i in range(1, n+1):
                conv = nn.Sequential(nn.Conv2d(in_channels, out_
channels, kernel_size, stride, padding),
                                     nn.ReLU(inplace=True),)
                setattr(self, 'conv%d'%i, conv)
                in_channels = out_channels
        # initialise the blocks
        for m in self.children():
            init_weights(m, init_type='kaiming')
    def forward(self, inputs):
        x = inputs
        for i in range(1, self.n+1):
            conv = getattr(self, 'conv%d'%i)
            x = conv(x)
        return x
class UnetUp(nn.Module):
      def __init__(self,in_channels, out_channels, is_deconv, n_
concat=2):
        super(UnetUp, self).__init__()
             self.conv = UnetConv(in_channels+(n_concat-2)* out_
channels, out_channels, False)
        if is_deconv:
          self.up = nn.ConvTranspose2d(in_channels, out_channels,
kernel_size=4, stride=2, padding=1)
        else:
            self.up = nn.UpsamplingBilinear2d(scale_factor=2)
        # initialise the blocks
        for m in self.children():
          if m.__class__.__name__.find('UnetConv') != -1: continue
            init_weights(m, init_type='kaiming')
    def forward(self, inputs0,*input):
        outputs0 = self.up(inputs0)
        for i in range(len(input)):
            outputs0 = torch.cat([outputs0,input[i]], 1)
        return self.conv(outputs0)
class UnetUp4(nn.Module):
      def __init__(self,in_channels, out_channels, is_deconv, n_
concat=2):
        super(UnetUp4, self).__init__()
             self.conv = UnetConv(in_channels+(n_concat-2)* out_
channels, out_channels, False)
        if is_deconv:
          self.up = nn.ConvTranspose2d(in_channels, out_channels,
kernel_size=6, stride=4, padding=1)
        else:
            self.up = nn.UpsamplingBilinear2d(scale_factor=4)
        # initialise the blocks
```

```
        for m in self.children():
          if m.__class__.__name__.find('UnetConv') != -1: continue
            init_weights(m, init_type='kaiming')
    def forward(self, inputs0,*input):
        outputs0 = self.up(inputs0)
        for i in range(len(input)):
            outputs0 = torch.cat([outputs0,input[i]], 1)
        return self.conv(outputs0)
class GCN(nn.Module):
    def __init__(self, num_state, num_node, bias=False):
        super(GCN, self).__init__()
        self.conv1 = nn.Conv1d(num_node, num_node, kernel_size=1,
padding=0,
                    stride=1, groups=1, bias=True)
        self.relu = nn.LeakyReLU(0.2,inplace=True)
      self.conv2 = nn.Conv1d(num_state, num_state, kernel_size=1,
padding=0,
                    stride=1, groups=1, bias=bias)
    def forward(self, x):
       h = self.conv1(x.permute(0, 2, 1).contiguous()).permute(0,
2, 1)
        h = h + x
        h = self.relu(h)
        h = self.conv2(h)
        return h
class GloRe_Unit(nn.Module):
    def __init__(self, num_in, num_mid, stride=(1,1), kernel=1):
        super(GloRe_Unit, self).__init__()
        self.num_s = int(2 * num_mid)
        self.num_n = int(1 * num_mid)
        kernel_size = (kernel, kernel)
        padding = (1, 1) if kernel == 3 else (0, 0)
        # reduce dimension
       self.conv_state = Basconv(num_in, self.num_s, is_batchnorm
= True, kernel_size=kernel_size, padding=padding)
        # generate projection and inverse projection functions
        self.conv_proj = Basconv(num_in, self.num_n, is_batchnorm
= True,kernel_size=kernel_size, padding=padding)
      self.conv_reproj = Basconv(num_in, self.num_n, is_batchnorm
= True,kernel_size=kernel_size, padding=padding)
        # reasoning by graph convolution
        self.gcn1 = GCN(num_state=self.num_s, num_node=self.num_
n)
        self.gcn2 = GCN(num_state=self.num_s, num_node=self.num_
n)
        # fusion
              self.fc_2 = nn.Conv2d(self.num_s, num_in, kernel_
size=kernel_size, padding=padding, stride=(1,1),
                    groups=1, bias=False)
        self.blocker = nn.BatchNorm2d(num_in)
    def forward(self, x):
        batch_size = x.size(0)
        # generate projection and inverse projection matrices
```

```
        x_state_reshaped = self.conv_state(x).view(batch_size, 
self.num_s, -1)
        x_proj_reshaped = self.conv_proj(x).view(batch_size, 
self.num_n, -1)
        x_rproj_reshaped = self.conv_reproj(x).view(batch_size, 
self.num_n, -1)
        # project to node space
        x_n_state1 = torch.bmm(x_state_reshaped, x_proj_
reshaped.permute(0, 2, 1))
        x_n_state2 = x_n_state1 * (1. / x_state_reshaped.size(2))
        # graph convolution
        x_n_rel1 = self.gcn1(x_n_state2)
        x_n_rel2 = self.gcn2(x_n_rel1)
        # inverse project to original space
        x_state_reshaped = torch.bmm(x_n_rel2, x_rproj_reshaped)
        x_state = x_state_reshaped.view(batch_size, self.num_s, 
*x.size()[2:])
        # fusion
        out = x + self.blocker(self.fc_2(x_state))
        return out
    def img2df(img, mask):
        img[mask == 0] = 0
        img[mask == 2] = 0
        return img
    def feature_fusion(out1, out2):
        output2 = F.log_softmax(out2, dim=1)
        out1_bg = torch.zeros([out1.shape[0], 1, out1.shape[2], 
out1.shape[3]]).cuda()
        out1_disc = torch.zeros([out1.shape[0], 1, out1.shape[2], 
out1.shape[3]]).cuda()
        out2_layer = torch.zeros([out2.shape[0], 9, out2.shape[2], 
out2.shape[3]]).cuda()
        out1_bg[:, 0, :, :] = out1[:, 0, :, :]
        out1_disc[:, 0, :, :] = out1[:, 2, :, :]
        out2_layer[:, :, :, :] = out2[:, 1:, :, :]
        out = torch.cat([out1_bg, out2_layer, out1_disc], 1)
        return output2, out
```

## [Python3.9] Init_weights

```
import torch
import torch.nn as nn
from torch.nn import init
def weights_init_normal(m):
    classname = m.__class__.__name__
    #print(classname)
    if classname.find('Conv') != -1:
        init.normal_(m.weight.data, 0.0, 0.02)
    elif classname.find('Linear') != -1:
        init.normal_(m.weight.data, 0.0, 0.02)
    elif classname.find('BatchNorm') != -1:
        init.normal_(m.weight.data, 1.0, 0.02)
        init.constant_(m.bias.data, 0.0)
```

```
def weights_init_xavier(m):
    classname = m.__class__.__name__
    #print(classname)
    if classname.find('Conv') != -1:
        init.xavier_normal_(m.weight.data, gain=1)
    elif classname.find('Linear') != -1:
        init.xavier_normal_(m.weight.data, gain=1)
    elif classname.find('BatchNorm') != -1:
        init.normal_(m.weight.data, 1.0, 0.02)
        init.constant_(m.bias.data, 0.0)
def weights_init_kaiming(m):
    classname = m.__class__.__name__
    #print(classname)
    if classname.find('Conv') != -1:
        init.kaiming_normal_(m.weight.data, a=0, mode='fan_in')
    elif classname.find('Linear') != -1:
        init.kaiming_normal_(m.weight.data, a=0, mode='fan_in')
    elif classname.find('BatchNorm') != -1:
        init.normal_(m.weight.data, 1.0, 0.02)
        init.constant_(m.bias.data, 0.0)
def weights_init_orthogonal(m):
    classname = m.__class__.__name__
    #print(classname)
    if classname.find('Conv') != -1:
        init.orthogonal_(m.weight.data, gain=1)
    elif classname.find('Linear') != -1:
        init.orthogonal_(m.weight.data, gain=1)
    elif classname.find('BatchNorm') != -1:
        init.normal_(m.weight.data, 1.0, 0.02)
        init.constant_(m.bias.data, 0.0)
def init_weights(net, init_type='normal'):
    #print('initialization method [%s]' % init_type)
    if init_type == 'normal':
        net.apply(weights_init_normal)
    elif init_type == 'xavier':
        net.apply(weights_init_xavier)
    elif init_type == 'kaiming':
        net.apply(weights_init_kaiming)
    elif init_type == 'orthogonal':
        net.apply(weights_init_orthogonal)
    else:
         raise NotImplementedError('initialization method [%s] is
not implemented' % init_type)
```

### [Python3.9] Net_builder

```
from models.nets.TSNet import TSMGUNet
from models.nets.OSNet import OSMGUNet
def net_builder(name,pretrained_model=None, pretrained=False):
    if name == 'tsmgunet':
        net = TSMGUNet()
    elif name == 'osmgunet':
        net = OSMGUNet()
```

```
    else:
        raise NameError("Unknow Model Name!")
    return net
```

**[Python3.9] Seg_dataset**

```
from torch.utils.data import Dataset
from os.path import join, exists
from PIL import Image
import torch
import os
import os.path as osp
import numpy as np
import torchvision.transforms as tt
import data.seg_transforms as st
import PIL
import random
class segList(Dataset):
    def __init__(self, data_dir, phase, transforms):
        self.data_dir = data_dir
        self.phase = phase
        self.transforms = transforms
        self.image_list = None
        self.label_list = None
        self.bbox_list = None
        self.read_lists()
    def __getitem__(self, index):
        if self.phase == 'train':
              self.image_list = get_list_dir(self.phase, 'img',
self.data_dir)
             self.label_list = get_list_dir(self.phase, 'mask',
self.data_dir)
            data = [Image.open(self.image_list[index])]
            data.append(Image.open(self.label_list[index]))
            data = list(self.transforms(*data))
            data = [data[0],data[1].long()]
            return tuple(data)
        if self.phase == 'eval' or 'test':
              self.image_list = get_list_dir(self.phase, 'img',
self.data_dir)
             self.label_list = get_list_dir(self.phase, 'mask',
self.data_dir)
            data = [Image.open(self.image_list[index])]
            imt = torch.from_numpy(np.array(data[0]))
            data.append(Image.open(self.label_list[index]))
            data = list(self.transforms(*data))
            image = data[0]
            label = data[1]
            imn = self.image_list[index].split('/')[-1]
            return (image,label.long(),imt,imn)
        if self.phase == 'predict':
              self.image_list = get_list_dir(self.phase, 'img',
self.data_dir)
```

```
            data = [Image.open(self.image_list[index])]
            imt = torch.from_numpy(np.array(data[0]))
            data = list(self.transforms(*data))
            image = data[0]
            imn = self.image_list[index].split('/')[-1]
            return (image,imt,imn)
    def __len__(self):
        return len(self.image_list)
    def read_lists(self):
      self.image_list = get_list_dir(self.phase, 'img', self.data_
dir)
        print('Total amount of {} images is : {}'.format(self.phase,
len(self.image_list)))
def get_list_dir(phase, type, data_dir):
    data_dir = osp.join(data_dir, phase, type)
    files = os.listdir(data_dir)
    list_dir = []
    for file in files:
        file_dir = osp.join(data_dir, file)
        list_dir.append(file_dir)
    return list_dir
```

**[Python3.9] Seg_transforms**

```
import numbers
import random
import numpy as np
from PIL import Image, ImageOps
import torch
class Label_Transform(object):
    def __init__(self,label_pixel=(26, 51, 77, 102, 128, 153, 179,
204, 230, 255)):
        self.label_pixel = label_pixel
    def __call__(self, image, label, *args):
        label = np.array(label)
        for i in range(len(self.label_pixel)):
            label[label == self.label_pixel[i]] = i+1
        return image,Image.fromarray(label)
class Normalize(object):
    """Given mean: (R, G, B) and std: (R, G, B),
    will normalize each channel of the torch.*Tensor, i.e.
    channel = (channel - mean) / std
    """
    def __init__(self, mean, std):
        self.mean = torch.FloatTensor(mean)
        self.std = torch.FloatTensor(std)
    def __call__(self, image, label=None):
        for t, m, s in zip(image, self.mean, self.std):
            t.sub_(m).div_(s)
        if label is None:
            return image,
        else:
            return image, label
```

```
class ToTensor(object):
    """Converts a PIL.Image or numpy.ndarray (H x W x C) in the range
    [0, 255] to a torch.FloatTensor of shape (C x H x W) in the range
[0.0, 1.0].
    """
    def __call__(self, pic, label=None):
        if isinstance(pic, np.ndarray):
            # handle numpy array
            img = torch.from_numpy(pic)
        else:
            # handle PIL Image
                    img = torch.ByteTensor(torch.ByteStorage.from_
buffer(pic.tobytes()))
            # PIL image mode: 1, L, P, I, F, RGB, YCbCr, RGBA, CMYK
            if pic.mode == 'YCbCr':
                nchannel = 3
            else:
                nchannel = len(pic.mode)
            img = img.view(pic.size[1], pic.size[0], nchannel)
            # put it from HWC to CHW format
           # yikes, this transpose takes 80% of the loading time/CPU
           img = img.transpose(0, 1).transpose(0, 2).contiguous()
        img = img.float().div(255)
        if label is None:
            return img,
        else:
                     return img, torch.LongTensor(np.array(label,
dtype=np.int))
class Compose(object):
    """Composes several transforms together.
    """
    def __init__(self, transforms):
        self.transforms = transforms
    def __call__(self, *args):
        for t in self.transforms:
            args = t(*args)
        return args
```

## [Python3.9] Logger and Progress Bar

```
from __future__ import absolute_import
import os
import sys
import time
import torch.nn as nn
import torch.nn.init as init
# __all__ = ['Logger', "progress_bar"]
class Logger(object):
    "'Save training process to log file."'
    def __init__(self, fpath, title=None, resume=False):
        self.file = None
        self.resume = resume
        self.title = '' if title is None else title
```

```
        if fpath is not None:
            if resume:
                self.file = open(fpath, 'r')
                name = self.file.readline()
                self.names = name.rstrip().split('\t')
                self.numbers = {}
                for _, name in enumerate(self.names):
                    self.numbers[name] = []
                for numbers in self.file:
                    numbers = numbers.rstrip().split('\t')
                    for i in range(0, len(numbers)):
                   self.numbers[self.names[i]].append(numbers[i])
                self.file.close()
                self.file = open(fpath, 'a')
            else: # build a file
                self.file = open(fpath, 'w')
    def set_names(self, names): # names for every line
        if self.resume:
            pass
        # initialize numbers as empty list
        self.numbers = {}
        self.names = names
        for _, name in enumerate(self.names):
            self.file.write(name)
            self.file.write('\t')
            self.numbers[name] = []
        self.file.write('\n')
        self.file.flush()
    def append(self, numbers):
          assert len(self.names) == len(numbers), 'Numbers do not
match names'
        for index, num in enumerate(numbers):
            self.file.write("{0:6f}".format(num))
            self.file.write('\t')
            self.numbers[self.names[index]].append(num)
        self.file.write('\n')
        self.file.flush()
    def write(self, content):
        self.file.write(content)
        self.file.write('\n')
    def close(self):
        if self.file is not None:
            self.file.close()
```

### [Python3.9] Loss

```
import torch
import torch.nn as nn
import torch.nn.functional as F
from torch.autograd import Variable
import numpy as np
def loss_builder1():
    criterion_1_1 = nn.NLLLoss(ignore_index=255)
```

```
    criterion_1_2 = DiceLoss(class_num=3)
    criterion = [criterion_1_1,criterion_1_2]
    return criterion
def loss_builder2():
    criterion_2_1 = nn.NLLLoss(ignore_index=255)
    criterion_2_2 = DiceLoss(class_num=11)
    criterion = [criterion_2_1, criterion_2_2]
    return criterion
class DiceLoss(nn.Module):
    def __init__(self, class_num=11,smooth=1):
        super(DiceLoss, self).__init__()
        self.smooth = smooth
        self.class_num = class_num
    def forward(self, input, target):
        input = torch.exp(input)
        self.smooth = 0.
        Dice = Variable(torch.Tensor([0]).float()).cuda()
        for i in range(1,self.class_num):
            input_i = input[:,i,:,:]
            target_i = (target == i).float()
            intersect = (input_i*target_i).sum()
            union = torch.sum(input_i) + torch.sum(target_i)
            if target_i.sum() == 0:
                dice = Variable(torch.Tensor([1]).float()).cuda()
            else:
                    dice = (2 * intersect + self.smooth) / (union +
self.smooth)
            Dice += dice
        dice_loss = 1 - Dice/(self.class_num - 1)
        return dice_loss
```

## [Python3.9] Utils_utils

```
import numpy as np
import torch
import os
import os.path as osp
import cv2
import scipy.misc as misc
import shutil
from skimage import measure
import math
import traceback
from sklearn import metrics
import zipfile
def adjust_learning_rate(args, optimizer, epoch):
    """
    Sets the learning rate to the initial LR decayed by 10 every 30
epochs(step = 30)
    """
    if args.lr_mode == 'step':
        lr = args.lr * (0.1 ** (epoch // args.step))
    elif args.lr_mode == 'poly':
```

```
        lr = args.lr * (1 - epoch / args.epochs) ** 0.9
    else:
      raise ValueError('Unknown lr mode {}'.format(args.lr_mode))
    for param_group in optimizer.param_groups:
        param_group['lr'] = lr
    return lr
class AverageMeter(object):
    """Computes and stores the average and current value"""
    def __init__(self):
        self.reset()
    def reset(self):
        self.val = 0
        self.avg = 0
        self.sum = 0
        self.count = 0
    def update(self, val, n=1):
        self.val = val
        self.sum += val * n
        self.count += n
        self.avg = self.sum / self.count
def save_model(state, is_best, model_path):
           model_latest_path   =   osp.join(model_path,'model_
latest.pth.tar')
    torch.save(state, model_latest_path)
    if is_best:
     model_best_path = osp.join(model_path,'model_best.pth.tar')
        shutil.copyfile(model_latest_path, model_best_path)
def save_dice_single(is_best, filename='dice_single.txt'):
    if is_best:
        shutil.copyfile(filename, 'dice_best.txt')
def compute_dice(ground_truth, prediction):
    ground_truth = ground_truth.flatten()
    prediction = prediction.flatten()
    try:
       ret = [0.5, 0.5, 0.5, 0.5, 0.5, 0.5, 0.5, 0.5, 0.5, 0.5, 0.5]
        for i in range(11):
            mask1 = (ground_truth == i)
            mask2 = (prediction == i)
            if mask1.sum() != 0:
                   ret[i] = float(2 * ((mask1 * (ground_truth ==
prediction)).sum()) / (mask1.sum() + mask2.sum()))
            else:
                ret[i] = float('nan')
    except Exception as e:
        traceback.print_exc()
        print("ERROR msg:", e)
        return None
    return ret
def compute_pa(ground_truth, prediction):
    ground_truth = ground_truth.flatten()
    prediction = prediction.flatten()
    try:
       ret = [0.5, 0.5, 0.5, 0.5, 0.5, 0.5, 0.5, 0.5, 0.5, 0.5, 0.5]
        for i in range(11):
```

```
            mask1 = (ground_truth == i)
            if mask1.sum() != 0:
                  ret[i] = float(((mask1 * (ground_truth ==
prediction)).sum()) / (mask1.sum()))
            else:
                ret[i] = float('nan')
    except Exception as e:
        traceback.print_exc()
        print("ERROR msg:", e)
        return None
    return ret
def compute_avg_score(ret_seg):
     BG, NFL_seg, GCL_seg, IPL_seg, INL_seg, OPL_seg, ONL_seg, IS_
OS_seg, RPE_seg, Choroid_seg, Disc_seg = 0.0, 0.0, 0.0, 0.0, 0.0,
0.0, 0.0, 0.0, 0.0, 0.0, 0.0
    n0, n1, n2, n3, n4, n5, n6, n7, n8, n9, n10 = 0.0, 0.0, 0.0, 0.0,
0.0, 0.0, 0.0, 0.0, 0.0, 0.0, 0.000001
    num = np.array(ret_seg).shape[0]
    for i in range(num):
        if not math.isnan(ret_seg[i][0]):
            BG += ret_seg[i][0]
            n0 += 1
        if not math.isnan(ret_seg[i][1]):
            NFL_seg += ret_seg[i][1]
            n1 += 1
        if not math.isnan(ret_seg[i][2]):
            GCL_seg += ret_seg[i][2]
            n2 += 1
        if not math.isnan(ret_seg[i][3]):
            IPL_seg += ret_seg[i][3]
            n3 += 1
        if not math.isnan(ret_seg[i][4]):
            INL_seg += ret_seg[i][4]
            n4 += 1
        if not math.isnan(ret_seg[i][5]):
            OPL_seg += ret_seg[i][5]
            n5 += 1
        if not math.isnan(ret_seg[i][6]):
            ONL_seg += ret_seg[i][6]
            n6 += 1
        if not math.isnan(ret_seg[i][7]):
            IS_OS_seg += ret_seg[i][7]
            n7 += 1
        if not math.isnan(ret_seg[i][8]):
            RPE_seg += ret_seg[i][8]
            n8 += 1
        if not math.isnan(ret_seg[i][9]):
            Choroid_seg += ret_seg[i][9]
            n9 += 1
        if not math.isnan(ret_seg[i][10]):
            Disc_seg += ret_seg[i][10]
            n10 += 1
    BG /= n0
    NFL_seg /= n1
```

```
    GCL_seg /= n2
    IPL_seg /= n3
    INL_seg /= n4
    OPL_seg /= n5
    ONL_seg /= n6
    IS_OS_seg /= n7
    RPE_seg /= n8
    Choroid_seg /= n9
    Disc_seg /= n10
    avg_seg = (NFL_seg + GCL_seg + IPL_seg + INL_seg + OPL_seg +
ONL_seg + IS_OS_seg + RPE_seg + Choroid_seg + Disc_seg) / 10
    return avg_seg, NFL_seg, GCL_seg, IPL_seg, INL_seg, OPL_seg,
ONL_seg, IS_OS_seg, RPE_seg, Choroid_seg, Disc_seg
def compute_single_avg_score(ret_seg):
    NFL_seg, GCL_seg, IPL_seg, INL_seg, OPL_seg, ONL_seg, IS_OS_
seg, RPE_seg, Choroid_seg, Disc_seg = 0.0, 0.0, 0.0, 0.0, 0.0, 0.0,
0.0, 0.0, 0.0, 0.0
    if not math.isnan(ret_seg[1]):
        NFL_seg = ret_seg[1]
    if not math.isnan(ret_seg[2]):
        GCL_seg = ret_seg[2]
    if not math.isnan(ret_seg[3]):
        IPL_seg = ret_seg[3]
    if not math.isnan(ret_seg[4]):
        INL_seg = ret_seg[4]
    if not math.isnan(ret_seg[5]):
        OPL_seg = ret_seg[5]
    if not math.isnan(ret_seg[6]):
        ONL_seg = ret_seg[6]
    if not math.isnan(ret_seg[7]):
        IS_OS_seg = ret_seg[7]
    if not math.isnan(ret_seg[8]):
        RPE_seg = ret_seg[8]
    if not math.isnan(ret_seg[9]):
        Choroid_seg = ret_seg[9]
    if not math.isnan(ret_seg[10]):
        Disc_seg = ret_seg[10]
    avg_seg = (NFL_seg + GCL_seg + IPL_seg + INL_seg + OPL_seg +
ONL_seg + IS_OS_seg + RPE_seg + Choroid_seg + Disc_seg) / 10
    return avg_seg
```

**[Python3.9] Vis**

```
import numpy as np
import os
import os.path as osp
import cv2

def vis_result(imn, imt, ant, pred, save_dir, n_class=11):
    img = gray2rgbimage(imt)
    imn = imn[0]
    pred_img = draw_img(imt, pred, n_class=n_class)
    if(ant is None):
```

```
      cv2.imwrite(osp.join(save_dir, imn), np.hstack((img, pred_
img)).astype('uint8'))
    else:
        ant_img = draw_img(imt, ant, n_class=n_class)
        cv2.imwrite(osp.join(save_dir, imn), np.hstack((img,ant_
img,pred_img )).astype('uint8'))
            cv2.imwrite(osp.join(save_dir, 'label/' + imn), ant_
img.astype('uint8'))
            cv2.imwrite(osp.join(save_dir, 'pred/' + imn), pred_
img.astype('uint8'))
def draw_img(img, seg, title = None, n_class=11):
    mask = img
    label_set = [i+1 for i in range(n_class)]
    color_set = {
                #0:(200,200,200),
                1:(255, 0, 0), #BGR #NFL
                2:(0, 255, 0), #GCL
                3:(0, 0, 255), #IPL
                4:(0, 255, 255), #INL
                5: (255, 0, 255), #OPL
                6: (255, 255, 0), #ONL
                7: (0, 0, 150), #IS/OS
                8: (0, 150, 0), #RPE
                9: (150, 0, 150), #choroid
                10: (100, 50, 250), # choroid
                11: (50, 100, 250), # choroid
            }
    mask = gray2rgbimage(mask)
    img = gray2rgbimage(img)
    if(title is not None):
       mask = cv2.putText(mask, title, (16, 16), cv2.FONT_HERSHEY_
SIMPLEX, 0.5, (255, 255, 255), 1) # white title
    for draw_label in label_set:
        mask[:, :, 0][seg[0,:,:] == draw_label] =(color_set[draw_
label][0])
       mask[:, :, 1][seg[0,:,:] == draw_label] = ( color_set[draw_
label][1])
       mask[:, :, 2][seg[0,:,:] == draw_label] = (color_set[draw_
label][2])
    img_mask = cv2.addWeighted(img,0.4,mask,0.6,0)
    return img_mask
def gray2rgbimage(image):
        a,b = image.shape
        new_img = np.ones((a,b,3))
        new_img[:,:,0] = image.reshape((a,b)).astype('uint8')
        new_img[:,:,1] = image.reshape((a,b)).astype('uint8')
        new_img[:,:,2] = image.reshape((a,b)).astype('uint8')
        return new_img
```

### [Python3.9] Main_os

```
##### System library #####
import os
```

```
os.environ["CUDA_VISIBLE_DEVICES"] = "2"
import os.path as osp
from os.path import exists
import argparse
import json
import logging
import time
import copy
##### pytorch library #####
import torch
from torch import nn
import torch.backends.cudnn as cudnn
from torch.autograd import Variable
##### My own library #####
import data.seg_transforms as dt
from data.seg_dataset import segList
from utils.logger import Logger
from models.net_builder import net_builder
from utils.loss import loss_builder1,loss_builder2
from utils.utils import adjust_learning_rate
from utils.utils import AverageMeter,save_model
from utils.utils import compute_dice,compute_pa,compute_single_
avg_score
from utils.vis import vis_result
# logger vis
FORMAT = "[%(asctime)-15s %(filename)s:%(lineno)d %(funcName)s]
%(message)s"
logging.basicConfig(format=FORMAT)
logger_vis = logging.getLogger(__name__)
logger_vis.setLevel(logging.DEBUG)
# training process
def   train(args,train_loader,   model,   criterion2,   opti-
mizer,epoch,print_freq=10):
    # set the AverageMeter
    batch_time = AverageMeter()
    losses = AverageMeter()
    dice = AverageMeter()
                  Dice_1,   Dice_2,   Dice_3,   Dice_4,   Dice_5,
Dice_6,    Dice_7,    Dice_8,    Dice_9,    Dice_10    =    Aver-
ageMeter(),AverageMeter(),AverageMeter(),AverageMeter(),
AverageMeter(), AverageMeter(), AverageMeter(), AverageMeter(),
AverageMeter(), AverageMeter()
    # switch to train mode
    model.train()
    end = time.time()
    for i, (input, target) in enumerate(train_loader):
        # variable
        input_var = Variable(input).cuda()
        target_var_seg = Variable(target).cuda()
        input_var1 = copy.deepcopy(input_var)
        # forward
        output_seg = model(input_var1)
        # calculate loss
        loss_2_1 = criterion2[0](output_seg, target_var_seg)
```

```
        loss_2_2 = criterion2[1](output_seg, target_var_seg)
       loss_2= loss_2_1 + loss_2_2 # loss from the two-stage network
        loss = loss_2
        losses.update(loss.data, input.size(0))
        # calculate dice score for segmentation
        _, pred_seg = torch.max(output_seg, 1)
        pred_seg = pred_seg.cpu().data.numpy()
        label_seg = target_var_seg.cpu().data.numpy()
        ret_d = compute_dice(label_seg, pred_seg)
        dice_score = compute_single_avg_score(ret_d)
        # update dice score
        dice.update(dice_score)
        Dice_1.update(ret_d[1])
        Dice_2.update(ret_d[2])
        Dice_3.update(ret_d[3])
        Dice_4.update(ret_d[4])
        Dice_5.update(ret_d[5])
        Dice_6.update(ret_d[6])
        Dice_7.update(ret_d[7])
        Dice_8.update(ret_d[8])
        Dice_9.update(ret_d[9])
        Dice_10.update(ret_d[10])
        # backwards
        optimizer.zero_grad()
        loss.backward()
        optimizer.step()
        # measure elapsed time
        batch_time.update(time.time() - end)
        end = time.time()
        # logger vis
        if i % print_freq == 0:
            logger_vis.info('Epoch: [{0}][{1}/{2}]\t'
              'Time {batch_time.val:.3f} ({batch_time.avg:.3f})\t'
                    'Dice {dice.val:.4f} ({dice.avg:.4f})\t'
                    'Dice_1 {dice_1.val:.4f} ({dice_1.avg:.4f})\t'
                    'Dice_2 {dice_2.val:.4f} ({dice_2.avg:.4f})\t'
                    'Dice_3 {dice_3.val:.4f} ({dice_3.avg:.4f})\t'
                    'Dice_4 {dice_4.val:.4f} ({dice_4.avg:.4f})\t'
                    'Dice_5 {dice_5.val:.4f} ({dice_5.avg:.4f})\t'
                    'Dice_6 {dice_6.val:.4f} ({dice_6.avg:.4f})\t'
                    'Dice_7 {dice_7.val:.4f} ({dice_7.avg:.4f})\t'
                    'Dice_8 {dice_8.val:.4f} ({dice_8.avg:.4f})\t'
                    'Dice_9 {dice_9.val:.4f} ({dice_9.avg:.4f})\t'
                                'Dice_10 {dice_10.val:.4f} ({dice_
10.avg:.4f})\t'.format(
           epoch, i, len(train_loader), batch_time=batch_time,dice
=   dice,dice_1=Dice_1,dice_2=Dice_2,dice_3=Dice_3,dice_4=Dice_
4,dice_5=Dice_5,dice_6=Dice_6,dice_7=Dice_7,dice_8=Dice_
8,dice_9=Dice_9,dice_10=Dice_10))
            print('Loss :',loss.cpu().data.numpy())
        return  losses.avg,dice.avg,Dice_1.avg,Dice_2.avg,Dice_
3.avg,Dice_4.avg,Dice_5.avg,Dice_6.avg,Dice_7.avg,Dice_
8.avg,Dice_9.avg,Dice_10.avg
# evaluation process
```

```
def eval(phase, args, eval_data_loader, model,result_path = None,
logger = None):
    # set the AverageMeter
    batch_time = AverageMeter()
    dice = AverageMeter()
    mpa = AverageMeter()
              Dice_1,   Dice_2,   Dice_3,   Dice_4,   Dice_5,
Dice_6,   Dice_7,   Dice_8,   Dice_9,   Dice_10   =   Aver-
ageMeter(),AverageMeter(),AverageMeter(),AverageMeter(),
AverageMeter(), AverageMeter(), AverageMeter(), AverageMeter(),
AverageMeter(), AverageMeter()
     pa_1, pa_2, pa_3, pa_4, pa_5, pa_6, pa_7, pa_8, pa_9, pa_10
= AverageMeter(), AverageMeter(), AverageMeter(), AverageMeter(),
AverageMeter(), AverageMeter(), AverageMeter(), AverageMeter(),
AverageMeter(), AverageMeter()
    dice_list, mpa_list = [], []
    ret_dice, ret_pa = [], []
    # switch to eval mode
    model.eval()
    end = time.time()
    pred_seg_batch = []
    label_seg_batch = []
      for iter, (image, label, imt, imn) in enumerate(eval_data_
loader):
        with torch.no_grad():
            image_var = Variable(image).cuda()
            # model forward
            output_seg = model(image_var)
            _, pred_seg = torch.max(output_seg, 1)
            # save visualized result
           pred_seg = pred_seg.cpu().data.numpy().astype('uint8')
            if phase == 'eval' or phase == 'test':
                imt = (imt.squeeze().numpy()).astype('uint8')
                ant = label.numpy().astype('uint8')
                save_dir = osp.join(result_path, 'vis')
                if not exists(save_dir): os.makedirs(save_dir)
               if not exists(save_dir+'/label'):os.makedirs(save_
dir+'/label')
              if not exists(save_dir + '/pred'): os.makedirs(save_
dir + '/pred')
                vis_result(imn, imt, ant, pred_seg, save_dir)
                print('Saved visualized results!')
            # calculate dice and pa score for segmentation
            label_seg = label.numpy().astype('uint8')
            pred_seg_batch.append(pred_seg)
            label_seg_batch.append(label_seg)
            ret_d = compute_dice(label_seg, pred_seg)
            ret_p = compute_pa(label_seg, pred_seg)
            ret_dice.append(ret_d)
            ret_pa.append(ret_p)
            dice_score = compute_single_avg_score(ret_d)
            mpa_score = compute_single_avg_score(ret_p)
            dice_list.append(dice_score)
            # update dice and pa score
```

```
            dice.update(dice_score)
            Dice_1.update(ret_d[1])
            Dice_2.update(ret_d[2])
            Dice_3.update(ret_d[3])
            Dice_4.update(ret_d[4])
            Dice_5.update(ret_d[5])
            Dice_6.update(ret_d[6])
            Dice_7.update(ret_d[7])
            Dice_8.update(ret_d[8])
            Dice_9.update(ret_d[9])
            Dice_10.update(ret_d[10])
            mpa_list.append(mpa_score)
            mpa.update(mpa_score)
            pa_1.update(ret_p[1])
            pa_2.update(ret_p[2])
            pa_3.update(ret_p[3])
            pa_4.update(ret_p[4])
            pa_5.update(ret_p[5])
            pa_6.update(ret_p[6])
            pa_7.update(ret_p[7])
            pa_8.update(ret_p[8])
            pa_9.update(ret_p[9])
            pa_10.update(ret_p[10])
        # measure elapsed time
        batch_time.update(time.time() - end)
        end = time.time()
        logger_vis.info('{0}: [{1}/{2}]\t'
                    'ID {id}\t'
                    'Dice {dice.val:.4f}\t'
                    'Dice_1 {dice_1.val:.4f}\t'
                    'Dice_2 {dice_2.val:.4f}\t'
                    'Dice_3 {dice_3.val:.4f}\t'
                    'Dice_4 {dice_4.val:.4f}\t'
                    'Dice_5 {dice_5.val:.4f}\t'
                    'Dice_6 {dice_6.val:.4f}\t'
                    'Dice_7 {dice_7.val:.4f}\t'
                    'Dice_8 {dice_8.val:.4f}\t'
                    'Dice_9 {dice_9.val:.4f}\t'
                    'Dice_10 {dice_10.val:.4f}\t'
                    'MPA {mpa.val:.4f}\t'
                    'PA_1 {pa_1.val:.4f}\t'
                    'PA_2 {pa_2.val:.4f}\t'
                    'PA_3 {pa_3.val:.4f}\t'
                    'PA_4 {pa_4.val:.4f}\t'
                    'PA_5 {pa_5.val:.4f}\t'
                    'PA_6 {pa_6.val:.4f}\t'
                    'PA_7 {pa_7.val:.4f}\t'
                    'PA_8 {pa_8.val:.4f}\t'
                    'PA_9 {pa_9.val:.4f}\t'
                    'PA_10 {pa_10.val:.4f}\t'
                    'Batch_time {batch_time.val:.3f}\t'
                      .format(phase.upper(), iter, len(eval_data_
loader),id=imn[0].split('.')[0], dice=dice, dice_1=Dice_1, dice_
2=Dice_2, dice_3=Dice_3,
```

```
                              dice_4=Dice_4, dice_5=Dice_5, dice_6=Dice_
6, dice_7=Dice_7, dice_8=Dice_8,
                                dice_9=Dice_9, dice_10=Dice_10, mpa=mpa,
pa_1=pa_1, pa_2=pa_2, pa_3=pa_3,
                               pa_4=pa_4, pa_5=pa_5, pa_6=pa_6, pa_7=pa_
7, pa_8=pa_8,
                               pa_9=pa_9, pa_10=pa_10, batch_time=batch_
time))
    # print final all dice and pa score
      final_dice_avg, final_dice_1, final_dice_2, final_dice_3,
final_dice_4, final_dice_5, final_dice_6, final_dice_7, final_
dice_8, final_dice_9, final_dice_10 = dice.avg, Dice_1.avg, Dice_
2.avg, Dice_3.avg, Dice_4.avg, Dice_5.avg, Dice_6.avg, Dice_
7.avg, Dice_8.avg, Dice_9.avg, Dice_10.avg
      final_pa_avg, final_pa_1, final_pa_2, final_pa_3, final_
pa_4, final_pa_5, final_pa_6, final_pa_7, final_pa_8, final_pa_9,
final_pa_10 = mpa.avg, pa_1.avg, pa_2.avg, pa_3.avg, pa_4.avg, pa_
5.avg, pa_6.avg, pa_7.avg, pa_8.avg, pa_9.avg, pa_10.avg
    print('###### Segmentation Result ######')
    print('Final Dice_avg Score:{:.4f}'.format(final_dice_avg))
    print('Final Dice_1 Score:{:.4f}'.format(final_dice_1))
    print('Final Dice_2 Score:{:.4f}'.format(final_dice_2))
    print('Final Dice_3 Score:{:.4f}'.format(final_dice_3))
    print('Final Dice_4 Score:{:.4f}'.format(final_dice_4))
    print('Final Dice_5 Score:{:.4f}'.format(final_dice_5))
    print('Final Dice_6 Score:{:.4f}'.format(final_dice_6))
    print('Final Dice_7 Score:{:.4f}'.format(final_dice_7))
    print('Final Dice_8 Score:{:.4f}'.format(final_dice_8))
    print('Final Dice_9 Score:{:.4f}'.format(final_dice_9))
    print('Final Dice_10 Score:{:.4f}'.format(final_dice_10))
    print('Final PA_avg:{:.4f}'.format(final_pa_avg))
    print('Final PA_1 Score:{:.4f}'.format(final_pa_1))
    print('Final PA_2 Score:{:.4f}'.format(final_pa_2))
    print('Final PA_3 Score:{:.4f}'.format(final_pa_3))
    print('Final PA_4 Score:{:.4f}'.format(final_pa_4))
    print('Final PA_5 Score:{:.4f}'.format(final_pa_5))
    print('Final PA_6 Score:{:.4f}'.format(final_pa_6))
    print('Final PA_7 Score:{:.4f}'.format(final_pa_7))
    print('Final PA_8 Score:{:.4f}'.format(final_pa_8))
    print('Final PA_9 Score:{:.4f}'.format(final_pa_9))
    print('Final PA_10 Score:{:.4f}'.format(final_pa_10))
    if phase == 'eval' or phase == 'test':
        logger.append(
        [final_dice_avg, final_dice_1, final_dice_2, final_dice_
3, final_dice_4, final_dice_5, final_dice_6, final_dice_7, final_
dice_8, final_dice_9, final_dice_10,
         final_pa_avg, final_pa_1, final_pa_2, final_pa_3, final_
pa_4, final_pa_5, final_pa_6, final_pa_7, final_pa_8, final_pa_9,
final_pa_10])
   return final_dice_avg, final_dice_1, final_dice_2, final_dice_
3, final_dice_4, final_dice_5, final_dice_6, final_dice_7, final_
dice_8, final_dice_9, final_dice_10,dice_list
###### train ######
def train_seg(args,train_result_path,train_loader,eval_loader):
```

```
    # logger setting
        logger_train = Logger(osp.join(train_result_path,'dice_
epoch.txt'), title='dice',resume=False)
logger_train.set_names(['Epoch','Dice_Train','Dice_Val','Dice_
1','Dice_11','Dice_2','Dice_22','Dice_3','Dice_33','Dice_
4','Dice_44','Dice_5','Dice_55','Dice_6','Dice_66','Dice_
7','Dice_77','Dice_8','Dice_88','Dice_9','Dice_99','Dice_
10','Dice_1010',])
    # print hyperparameters
    for k, v in args.__dict__.items():
        print(k, ':', v)
    # load the network
    net = net_builder(args.name)
    model = torch.nn.DataParallel(net).cuda()
    print('#'*15,args.name,'#'*15)
    # define loss function
    criterion2 = loss_builder2()
    # set optimizer
   optimizer = torch.optim.Adam(net.parameters(), #Adam optimizer
                              args.lr,
                              betas=(0.9, 0.99),
                              weight_decay=args.weight_decay)
    cudnn.benchmark = True
    # main training
    best_dice = 0
    start_epoch = 0
    for epoch in range(start_epoch, args.epochs):
        lr = adjust_learning_rate(args, optimizer, epoch)
       logger_vis.info('Epoch: [{0}]\tlr {1:.06f}'.format(epoch,
lr))
        # train for one epoch
                loss,dice_train,dice_1,dice_2,dice_3,dice_4,dice_
5,dice_6,dice_7,dice_8,dice_9,dice_10    =    train(args,train_
loader, model,criterion2, optimizer,epoch)
        # evaluate on validation set
dice_val,dice_11,dice_22,dice_33,dice_44,dice_55,dice_66,dice_
77,dice_88,dice_99,dice_1010,dice_list  =  eval('train',  args,
eval_loader, model)
        # save the best model
        is_best = dice_val > best_dice
        best_dice = max(dice_val, best_dice)
        model_dir = osp.join(train_result_path,'model')
        if not exists(model_dir):
            os.makedirs(model_dir)
        save_model({
            'epoch': epoch + 1,
            'state_dict': model.state_dict(),
            'dice_epoch':dice_val,
            'best_dice': best_dice,
        }, is_best, model_dir)
        # logger
logger_train.append([epoch,dice_train,dice_val,dice_1,dice_
11,dice_2,dice_22,dice_3,dice_33,dice_4,dice_44,dice_5,dice_
55,dice_6,dice_66,dice_7,dice_77,dice_8,dice_88,dice_9,dice_
99,dice_10,dice_1010])
```

```
###### validation ######
def eval_seg(args, eval_result_path, eval_loader):
    # logger setting
      logger_eval = Logger(osp.join(eval_result_path, 'dice_mpa_
epoch.txt'), title='dice&mpa', resume=False)
    logger_eval.set_names(
        ['Dice', 'Dice_1', 'Dice_2', 'Dice_3', 'Dice_4', 'Dice_5',
'Dice_6', 'Dice_7', 'Dice_8', 'Dice_9','Dice_10',
        'mpa', 'pa_1', 'pa_2','pa_3', 'pa_4', 'pa_5', 'pa_6', 'pa_
7', 'pa_8', 'pa_9','pa_10',])
    # load the model
    print('Loading eval model: {}'.format(args.name))
    net = net_builder(args.name)
    model = torch.nn.DataParallel(net).cuda()
    checkpoint = torch.load(args.model_path)
    model.load_state_dict(checkpoint['state_dict'])
    print('Model loaded!')
    cudnn.benchmark = True
    # evaluate the model on validation set
        eval('eval', args, eval_loader, model, eval_result_path,
logger_eval)
###### test ######
def test_seg(args, test_result_path, test_loader):
    # logger setting
      logger_test = Logger(osp.join(test_result_path, 'dice_mpa_
epoch.txt'), title='dice&mpa', resume=False)
    logger_test.set_names(
        ['Dice', 'Dice_1', 'Dice_2', 'Dice_3', 'Dice_4', 'Dice_5',
'Dice_6', 'Dice_7', 'Dice_8', 'Dice_9','Dice_10',
        'mpa', 'pa_1', 'pa_2','pa_3', 'pa_4', 'pa_5', 'pa_6', 'pa_
7', 'pa_8', 'pa_9','pa_10',])
# load the model
    print('Loading test model ...')
    net = net_builder(args.name)
    model = torch.nn.DataParallel(net).cuda()
    checkpoint = torch.load(args.model_path)
    model.load_state_dict(checkpoint['state_dict'])
    print('Model loaded!')
    cudnn.benchmark = True
    # test the model on testing set
        eval('test', args, test_loader, model, test_result_path,
logger_test)
def parse_args():
    parser = argparse.ArgumentParser(description='train')
    # config
          parser.add_argument('-d', '--data-dir', default=None,
required=True)
          parser.add_argument('--name', dest='name',help='change
model',default=None, type=str)
    parser.add_argument('-j', '--workers', type=int, default=2)
    # train setting
    parser.add_argument('--step', type=int, default=20)
        parser.add_argument('--batch-size', type=int, default=1,
metavar='N',
```

```
                help='input batch size for training (default: 1)')
        parser.add_argument('--epochs', type=int, default=10,
metavar='N',
                 help='number of epochs to train (default: 10)')
        parser.add_argument('--lr', type=float, default=0.01,
metavar='LR',
                 help='learning rate (default: 0.01)')
    parser.add_argument('--lr-mode', type=str, default='step')
     parser.add_argument('--momentum', type=float, default=0.9,
metavar='M',
                 help='SGD momentum (default: 0.9)')
    parser.add_argument('--weight-decay', '--wd', default=1e-4,
type=float,
                metavar='W', help='weight decay (default: 1e-4)')
    parser.add_argument('--t', type=str, default='t1')
     parser.add_argument('--model-path', help='pretrained model
test', default=' ', type=str)
    args = parser.parse_args()
    return args
def main():
    ##### config #####
    args = parse_args()
    seed = 1234
    torch.manual_seed(seed)
    torch.cuda.manual_seed(seed)
    print('torch version:',torch.__version__)
    ##### result path setting #####
    tn = args.t
    task_name = args.data_dir.split('/')[-2] + '/' + args.data_
dir.split('/')[-1]
                train_result_path   =   osp.join('result',task_
name,'train',args.name + '_' +str(args.lr) + '_'+ tn)
    if not exists(train_result_path):
        os.makedirs(train_result_path)
                 test_result_path   =   osp.join('result',task_
name,'test',args.name + '_' +str(args.lr) + '_'+ tn)
    if not exists(test_result_path):
        os.makedirs(test_result_path)
    ##### load dataset #####
    info = json.load(open(osp.join(args.data_dir, 'info.json'),
'r'))
    normalize = dt.Normalize(mean=info['mean'], std=info['std'])
    t = []
    t.extend([dt.Label_Transform(),dt.ToTensor(),normalize])
   train_dataset = segList(args.data_dir, 'train', dt.Compose(t))
    val_dataset = segList(args.data_dir, 'eval', dt.Compose(t))
    test_dataset = segList(args.data_dir, 'test', dt.Compose(t))
            train_loader  =  torch.utils.data.DataLoader(train_
dataset,batch_size=args.batch_size,     shuffle=True,     num_
workers=args.workers, pin_memory=True, drop_last=True)
        eval_loader = torch.utils.data.DataLoader(val_dataset,
batch_size=1,  shuffle=False,  num_workers=args.workers,  pin_
memory=False)
```

```
        test_loader = torch.utils.data.DataLoader(test_dataset,
batch_size=1, shuffle=False, num_workers=args.workers, pin_
memory=False)
    ##### train #####
    train_seg(args,train_result_path,train_loader,eval_loader)
    ##### test #####
            model_best_path = osp.join(osp.join(train_result_
path,'model'),'model_best.pth.tar')
    args.model_path = model_best_path
    test_seg(args,test_result_path,test_loader)
if __name__ == '__main__':
    main()
```

## [Python3.9] Main_ts

```
##### System library #####
import os
os.environ["CUDA_VISIBLE_DEVICES"] = "0"
import os
os.environ["KMP_DUPLICATE_LIB_OK"] = "TRUE"
import os.path as osp
from os.path import exists
import argparse
import json
import logging
import time
import copy
##### pytorch library #####
import torch
from torch import nn
import torch.backends.cudnn as cudnn
from torch.autograd import Variable
##### My own library #####
import data.seg_transforms as dt
from data.seg_dataset import segList
from models.net_builder import net_builder
from utils.logger import Logger
from utils.loss import loss_builder1,loss_builder2
from utils.utils import adjust_learning_rate
from utils.utils import AverageMeter,save_model
from utils.utils import compute_dice,compute_pa,compute_single_
avg_score
from utils.vis import vis_result
# logger vis
FORMAT = "[%(asctime)-15s %(filename)s:%(lineno)d %(funcName)s]
%(message)s"
logging.basicConfig(format=FORMAT)
logger_vis = logging.getLogger(__name__)
logger_vis.setLevel(logging.DEBUG)
# training process
def train(args,train_loader, model, criterion1, criterion2, opti-
mizer,epoch,print_freq=10):
    # set the AverageMeter
```

```
    batch_time = AverageMeter()
    losses = AverageMeter()
    dice = AverageMeter()
                 Dice_1,  Dice_2,  Dice_3,  Dice_4,  Dice_5,
Dice_6,  Dice_7,  Dice_8,  Dice_9,  Dice_10  =  Aver-
ageMeter(),AverageMeter(),AverageMeter(),AverageMeter(),
AverageMeter(), AverageMeter(), AverageMeter(), AverageMeter(),
AverageMeter(), AverageMeter()
    # switch to train mode
    model.train()
    end = time.time()
    for i, (input, target) in enumerate(train_loader):
        # variable
        input_var = Variable(input).cuda()
        target_var_seg = Variable(target).cuda()
        target_var_seg1 = copy.deepcopy(target_var_seg)
        input_var1 = copy.deepcopy(input_var)
        # forward
        output_seg1,_,output_seg = model(input_var1)
        # modify label for the first stage network
        target_var_seg1[target_var_seg==0]=0
        target_var_seg1[target_var_seg==1]=1
        target_var_seg1[target_var_seg==2]=1
        target_var_seg1[target_var_seg==3]=1
        target_var_seg1[target_var_seg==4]=1
        target_var_seg1[target_var_seg==5]=1
        target_var_seg1[target_var_seg==6]=1
        target_var_seg1[target_var_seg==7]=1
        target_var_seg1[target_var_seg==8]=1
        target_var_seg1[target_var_seg==9]=1
        target_var_seg1[target_var_seg==10] = 2
        # calculate loss
        loss_1_1 = criterion1[0](output_seg1, target_var_seg1)
        loss_1_2 = criterion1[1](output_seg1, target_var_seg1)
          loss_1 = loss_1_1 + loss_1_2 # loss from the first stage
network
        loss_2_1 = criterion2[0](output_seg, target_var_seg)
        loss_2_2 = criterion2[1](output_seg, target_var_seg)
       loss_2 = loss_2_1 + loss_2_2 # loss from the two-stage network
        loss = loss_1 + 2*loss_2
        losses.update(loss.data, input.size(0))
        # calculate dice score for segmentation
        _, pred_seg = torch.max(output_seg, 1)
        pred_seg = pred_seg.cpu().data.numpy()
        label_seg = target_var_seg.cpu().data.numpy()
        ret_d = compute_dice(label_seg, pred_seg)
        dice_score = compute_single_avg_score(ret_d)
        # update dice score
        dice.update(dice_score)
        Dice_1.update(ret_d[1])
        Dice_2.update(ret_d[2])
        Dice_3.update(ret_d[3])
        Dice_4.update(ret_d[4])
        Dice_5.update(ret_d[5])
```

```
        Dice_6.update(ret_d[6])
        Dice_7.update(ret_d[7])
        Dice_8.update(ret_d[8])
        Dice_9.update(ret_d[9])
        Dice_10.update(ret_d[10])
        # backwards
        optimizer.zero_grad()
        loss.backward()
        optimizer.step()
        # measure elapsed time
        batch_time.update(time.time() - end)
        end = time.time()
        # logger vis
        if i % print_freq == 0:
            logger_vis.info('Epoch: [{0}][{1}/{2}]\t'
             'Time {batch_time.val:.3f} ({batch_time.avg:.3f})\t'
                   'Dice {dice.val:.4f} ({dice.avg:.4f})\t'
                   'Dice_1 {dice_1.val:.4f} ({dice_1.avg:.4f})\t'
                   'Dice_2 {dice_2.val:.4f} ({dice_2.avg:.4f})\t'
                   'Dice_3 {dice_3.val:.4f} ({dice_3.avg:.4f})\t'
                   'Dice_4 {dice_4.val:.4f} ({dice_4.avg:.4f})\t'
                   'Dice_5 {dice_5.val:.4f} ({dice_5.avg:.4f})\t'
                   'Dice_6 {dice_6.val:.4f} ({dice_6.avg:.4f})\t'
                   'Dice_7 {dice_7.val:.4f} ({dice_7.avg:.4f})\t'
                   'Dice_8 {dice_8.val:.4f} ({dice_8.avg:.4f})\t'
                   'Dice_9 {dice_9.val:.4f} ({dice_9.avg:.4f})\t'
                              'Dice_10 {dice_10.val:.4f} ({dice_
10.avg:.4f})\t'.format(
         epoch, i, len(train_loader), batch_time=batch_time,dice
=  dice,dice_1=Dice_1,dice_2=Dice_2,dice_3=Dice_3,dice_4=Dice_
4,dice_5=Dice_5,dice_6=Dice_6,dice_7=Dice_7,dice_8=Dice_
8,dice_9=Dice_9,dice_10=Dice_10))
        print('Loss :',loss.cpu().data.numpy())
         return  losses.avg,dice.avg,Dice_1.avg,Dice_2.avg,Dice_
3.avg,Dice_4.avg,Dice_5.avg,Dice_6.avg,Dice_7.avg,Dice_
8.avg,Dice_9.avg,Dice_10.avg
# evaluation process
def eval(phase, args, eval_data_loader, model, result_path = None,
logger = None):
    # set the AverageMeter
    batch_time = AverageMeter()
    dice = AverageMeter()
    mpa = AverageMeter()
                Dice_1,   Dice_2,   Dice_3,   Dice_4,   Dice_5,
Dice_6,    Dice_7,    Dice_8,    Dice_9,    Dice_10    =    Aver-
ageMeter(),AverageMeter(),AverageMeter(),AverageMeter(),
AverageMeter(), AverageMeter(), AverageMeter(), AverageMeter(),
AverageMeter(), AverageMeter()
     pa_1, pa_2, pa_3, pa_4, pa_5, pa_6, pa_7, pa_8, pa_9, pa_10
=AverageMeter(),AverageMeter(),AverageMeter(),AverageMeter(),
AverageMeter(), AverageMeter(), AverageMeter(), AverageMeter(),
AverageMeter(), AverageMeter()
    dice_list, mpa_list = [], []
    ret_dice, ret_pa = [], []
```

```
    # switch to eval mode
    model.eval()
    end = time.time()
    pred_seg_batch = []
    label_seg_batch = []
      for iter, (image, label, imt, imn) in enumerate(eval_data_
loader):
        with torch.no_grad():
            image_var = Variable(image).cuda()
            # model forward
            _,_,output_seg = model(image_var)
            _, pred_seg = torch.max(output_seg, 1)
            # save visualized result
          pred_seg = pred_seg.cpu().data.numpy().astype('uint8')
            if phase == 'eval' or phase == 'test':
                imt = (imt.squeeze().numpy()).astype('uint8')
                ant = label.numpy().astype('uint8')
                save_dir = osp.join(result_path, 'vis')
                if not exists(save_dir): os.makedirs(save_dir)
               if not exists(save_dir+'/label'):os.makedirs(save_
dir+'/label')
              if not exists(save_dir + '/pred'): os.makedirs(save_
dir + '/pred')
                vis_result(imn, imt, ant, pred_seg, save_dir)
                print('Saved visualized results!')
            # calculate dice and pa score for segmentation
            label_seg = label.numpy().astype('uint8')
            pred_seg_batch.append(pred_seg)
            label_seg_batch.append(label_seg)
            ret_d = compute_dice(label_seg, pred_seg)
            ret_p = compute_pa(label_seg, pred_seg)
            ret_dice.append(ret_d)
            ret_pa.append(ret_p)
            dice_score = compute_single_avg_score(ret_d)
            mpa_score = compute_single_avg_score(ret_p)
            dice_list.append(dice_score)
            # update dice and pa score
            dice.update(dice_score)
            Dice_1.update(ret_d[1])
            Dice_2.update(ret_d[2])
            Dice_3.update(ret_d[3])
            Dice_4.update(ret_d[4])
            Dice_5.update(ret_d[5])
            Dice_6.update(ret_d[6])
            Dice_7.update(ret_d[7])
            Dice_8.update(ret_d[8])
            Dice_9.update(ret_d[9])
            Dice_10.update(ret_d[10])
            mpa_list.append(mpa_score)
            mpa.update(mpa_score)
            pa_1.update(ret_p[1])
            pa_2.update(ret_p[2])
            pa_3.update(ret_p[3])
            pa_4.update(ret_p[4])
```

```
            pa_5.update(ret_p[5])
            pa_6.update(ret_p[6])
            pa_7.update(ret_p[7])
            pa_8.update(ret_p[8])
            pa_9.update(ret_p[9])
            pa_10.update(ret_p[10])
        # measure elapsed time
        batch_time.update(time.time() - end)
        end = time.time()
        logger_vis.info('{0}: [{1}/{2}]\t'
                    'ID {id}\t'
                    'Dice {dice.val:.4f}\t'
                    'Dice_1 {dice_1.val:.4f}\t'
                    'Dice_2 {dice_2.val:.4f}\t'
                    'Dice_3 {dice_3.val:.4f}\t'
                    'Dice_4 {dice_4.val:.4f}\t'
                    'Dice_5 {dice_5.val:.4f}\t'
                    'Dice_6 {dice_6.val:.4f}\t'
                    'Dice_7 {dice_7.val:.4f}\t'
                    'Dice_8 {dice_8.val:.4f}\t'
                    'Dice_9 {dice_9.val:.4f}\t'
                    'Dice_10 {dice_10.val:.4f}\t'
                    'MPA {mpa.val:.4f}\t'
                    'PA_1 {pa_1.val:.4f}\t'
                    'PA_2 {pa_2.val:.4f}\t'
                    'PA_3 {pa_3.val:.4f}\t'
                    'PA_4 {pa_4.val:.4f}\t'
                    'PA_5 {pa_5.val:.4f}\t'
                    'PA_6 {pa_6.val:.4f}\t'
                    'PA_7 {pa_7.val:.4f}\t'
                    'PA_8 {pa_8.val:.4f}\t'
                    'PA_9 {pa_9.val:.4f}\t'
                    'PA_10 {pa_10.val:.4f}\t'
                    'Batch_time {batch_time.val:.3f}\t'
                      .format(phase.upper(), iter, len(eval_data_
loader),id=imn[0].split('.')[0], dice=dice, dice_1=Dice_1, dice_
2=Dice_2, dice_3=Dice_3,
                       dice_4=Dice_4, dice_5=Dice_5, dice_6=Dice_
6, dice_7=Dice_7, dice_8=Dice_8,
                         dice_9=Dice_9, dice_10=Dice_10, mpa=mpa,
pa_1=pa_1, pa_2=pa_2, pa_3=pa_3,
                         pa_4=pa_4, pa_5=pa_5, pa_6=pa_6, pa_7=pa_
7, pa_8=pa_8,
                        pa_9=pa_9, pa_10=pa_10, batch_time=batch_
time))
    # print final all dice and pa score
      final_dice_avg, final_dice_1, final_dice_2, final_dice_3,
final_dice_4, final_dice_5, final_dice_6, final_dice_7, final_
dice_8, final_dice_9, final_dice_10 = dice.avg, Dice_1.avg, Dice_
2.avg, Dice_3.avg, Dice_4.avg, Dice_5.avg, Dice_6.avg, Dice_
7.avg, Dice_8.avg, Dice_9.avg, Dice_10.avg
      final_pa_avg, final_pa_1, final_pa_2, final_pa_3, final_
pa_4, final_pa_5, final_pa_6, final_pa_7, final_pa_8, final_pa_9,
final_pa_10 = mpa.avg, pa_1.avg, pa_2.avg, pa_3.avg, pa_4.avg, pa_
5.avg, pa_6.avg, pa_7.avg, pa_8.avg, pa_9.avg, pa_10.avg
```

```
    print('###### Segmentation Result ######')
    print('Final Dice_avg Score:{:.4f}'.format(final_dice_avg))
    print('Final Dice_1 Score:{:.4f}'.format(final_dice_1))
    print('Final Dice_2 Score:{:.4f}'.format(final_dice_2))
    print('Final Dice_3 Score:{:.4f}'.format(final_dice_3))
    print('Final Dice_4 Score:{:.4f}'.format(final_dice_4))
    print('Final Dice_5 Score:{:.4f}'.format(final_dice_5))
    print('Final Dice_6 Score:{:.4f}'.format(final_dice_6))
    print('Final Dice_7 Score:{:.4f}'.format(final_dice_7))
    print('Final Dice_8 Score:{:.4f}'.format(final_dice_8))
    print('Final Dice_9 Score:{:.4f}'.format(final_dice_9))
    print('Final Dice_10 Score:{:.4f}'.format(final_dice_10))
    print('Final PA_avg:{:.4f}'.format(final_pa_avg))
    print('Final PA_1 Score:{:.4f}'.format(final_pa_1))
    print('Final PA_2 Score:{:.4f}'.format(final_pa_2))
    print('Final PA_3 Score:{:.4f}'.format(final_pa_3))
    print('Final PA_4 Score:{:.4f}'.format(final_pa_4))
    print('Final PA_5 Score:{:.4f}'.format(final_pa_5))
    print('Final PA_6 Score:{:.4f}'.format(final_pa_6))
    print('Final PA_7 Score:{:.4f}'.format(final_pa_7))
    print('Final PA_8 Score:{:.4f}'.format(final_pa_8))
    print('Final PA_9 Score:{:.4f}'.format(final_pa_9))
    print('Final PA_10 Score:{:.4f}'.format(final_pa_10))
    if phase == 'eval' or phase == 'test':
        logger.append(
        [final_dice_avg, final_dice_1, final_dice_2, final_dice_
3, final_dice_4, final_dice_5, final_dice_6, final_dice_7, final_
dice_8, final_dice_9, final_dice_10,
        final_pa_avg, final_pa_1, final_pa_2, final_pa_3, final_
pa_4, final_pa_5, final_pa_6, final_pa_7, final_pa_8, final_pa_9,
final_pa_10])
   return final_dice_avg, final_dice_1, final_dice_2, final_dice_
3, final_dice_4, final_dice_5, final_dice_6, final_dice_7, final_
dice_8, final_dice_9, final_dice_10,dice_list
###### train ######
def train_seg(args,train_result_path,train_loader,eval_loader):
    # logger setting
        logger_train = Logger(osp.join(train_result_path,'dice_
epoch.txt'), title='dice',resume=False)


logger_train.set_names(['Epoch','Dice_Train','Dice_Val','Dice_
1','Dice_11','Dice_2','Dice_22','Dice_3','Dice_33','Dice_
4','Dice_44','Dice_5','Dice_55','Dice_6','Dice_66','Dice_
7','Dice_77','Dice_8','Dice_88','Dice_9','Dice_99','Dice_
10','Dice_1010',])
    # print hyperparameters
    for k, v in args.__dict__.items():
        print(k, ':', v)
    # load the network
    net = net_builder(args.name)
    model = torch.nn.DataParallel(net).cuda()
    # model = torch.nn.DataParallel(net).cpu()
    print('#'*15,args.name,'#'*15)
```

```
    # define loss function
    criterion1 = loss_builder1()
    criterion2 = loss_builder2()
    # set optimizer
   optimizer = torch.optim.Adam(net.parameters(), #Adam optimizer
                          args.lr,
                          betas=(0.9, 0.99),
                          weight_decay=args.weight_decay)
    cudnn.benchmark = True
    # main training
    best_dice = 0
    start_epoch = 0
    for epoch in range(start_epoch, args.epochs):
        lr = adjust_learning_rate(args,optimizer, epoch)
        logger_vis.info('Epoch: [{0}]\t'.format(epoch))
        # train for one epoch
               loss,dice_train,dice_1,dice_2,dice_3,dice_4,dice_
5,dice_6,dice_7,dice_8,dice_9,dice_10    =    train(args,train_
loader, model,criterion1, criterion2, optimizer,epoch)
        # evaluate on validation set
dice_val,dice_11,dice_22,dice_33,dice_44,dice_55,dice_66,dice_
77,dice_88,dice_99,dice_1010,dice_list  =  eval('train',  args,
eval_loader, model,criterion1, criterion2, )
        # save the best model
        is_best = dice_val > best_dice
        best_dice = max(dice_val, best_dice)
        model_dir = osp.join(train_result_path,'model')
        if not exists(model_dir):
            os.makedirs(model_dir)
        save_model({
            'epoch': epoch + 1,
            'state_dict': model.state_dict(),
            'dice_epoch':dice_val,
            'best_dice': best_dice,
        }, is_best, model_dir)
        # logger
logger_train.append([epoch,dice_train,dice_val,dice_1,dice_
11,dice_2,dice_22,dice_3,dice_33,dice_4,dice_44,dice_5,dice_
55,dice_6,dice_66,dice_7,dice_77,dice_8,dice_88,dice_9,dice_
99,dice_10,dice_1010])
###### validation ######
def eval_seg(args, eval_result_path, eval_loader):
    # logger setting
      logger_eval = Logger(osp.join(eval_result_path, 'dice_mpa_
epoch.txt'), title='dice&mpa', resume=False)
    logger_eval.set_names(
        ['Dice', 'Dice_1', 'Dice_2', 'Dice_3', 'Dice_4', 'Dice_5',
'Dice_6', 'Dice_7', 'Dice_8', 'Dice_9','Dice_10',
        'mpa', 'pa_1', 'pa_2','pa_3', 'pa_4', 'pa_5', 'pa_6', 'pa_
7', 'pa_8', 'pa_9','pa_10',])
    # load the model
    print('Loading eval model: {}'.format(args.name))
    net = net_builder(args.name)
    model = torch.nn.DataParallel(net).cuda()
```

```
    checkpoint = torch.load(args.model_path)
    model.load_state_dict(checkpoint['state_dict'])
    print('Model loaded!')
    cudnn.benchmark = True
    # evaluate the model on validation set
       eval('eval', args, eval_loader, model, eval_result_path,
logger_eval)
###### test ######
def test_seg(args, test_result_path, test_loader):
    # logger setting
     logger_test = Logger(osp.join(test_result_path, 'dice_mpa_
epoch.txt'), title='dice&mpa', resume=False)
    logger_test.set_names(
        ['Dice', 'Dice_1', 'Dice_2', 'Dice_3', 'Dice_4', 'Dice_5',
'Dice_6', 'Dice_7', 'Dice_8', 'Dice_9','Dice_10',
        'mpa', 'pa_1', 'pa_2','pa_3', 'pa_4', 'pa_5', 'pa_6', 'pa_
7', 'pa_8', 'pa_9','pa_10',])
    # load the model
    print('Loading test model ...')
    net = net_builder(args.name)
    model = torch.nn.DataParallel(net).cuda()
    checkpoint = torch.load(args.model_path)
    model.load_state_dict(checkpoint['state_dict'])
    print('Model loaded!')
    cudnn.benchmark = True
    # test the model on testing set
       eval('test', args, test_loader, model, test_result_path,
logger_test)
def parse_args():
    parser = argparse.ArgumentParser(description='train')
    # config
          parser.add_argument('-d', '--data-dir', default=None,
required=True)
          parser.add_argument('--name', dest='name',help='change
model',default=None, type=str)
    parser.add_argument('-j', '--workers', type=int, default=1)
    # train setting
    parser.add_argument('--step', type=int, default=20)
       parser.add_argument('--batch-size', type=int, default=1,
metavar='N',
                  help='input batch size for training (default: 1)')
          parser.add_argument('--epochs', type=int, default=10,
metavar='N',
                    help='number of epochs to train (default: 10)')
          parser.add_argument('--lr', type=float, default=0.01,
metavar='LR',
                    help='learning rate (default: 0.01)')
    parser.add_argument('--lr-mode', type=str, default='step')
      parser.add_argument('--momentum', type=float, default=0.9,
metavar='M',
                    help='SGD momentum (default: 0.9)')
     parser.add_argument('--weight-decay', '--wd', default=1e-4,
type=float,
                  metavar='W', help='weight decay (default: 1e-4)')
```

```
    parser.add_argument('--t', type=str, default='t1')
      parser.add_argument('--model-path', help='pretrained model
test', default=' ', type=str)
    args = parser.parse_args()
    return args
def main():
    ##### config #####
    args = parse_args()
    seed = 1234
    torch.manual_seed(seed)
    torch.cuda.manual_seed(seed)
    print('torch version:',torch.__version__)
    ##### result path setting #####
    tn = args.t
     task_name = args.data_dir.split('/')[-2] + '/' + args.data_
dir.split('/')[-1]
                  train_result_path   =   osp.join('result',task_
name,'train',args.name + '_' +str(args.lr) + '_'+ tn)
    if not exists(train_result_path):
        os.makedirs(train_result_path)
                  test_result_path   =   osp.join('result',task_
name,'test',args.name + '_' +str(args.lr) + '_'+ tn)
    if not exists(test_result_path):
        os.makedirs(test_result_path)
    ##### load dataset #####
     info = json.load(open(osp.join(args.data_dir, 'info.json'),
'r'))
    normalize = dt.Normalize(mean=info['mean'], std=info['std'])
    t = []
    t.extend([dt.Label_Transform(),dt.ToTensor(),normalize])
   train_dataset = segList(args.data_dir, 'train', dt.Compose(t))
    val_dataset = segList(args.data_dir, 'eval', dt.Compose(t))
    test_dataset = segList(args.data_dir, 'test', dt.Compose(t))
            train_loader  =  torch.utils.data.DataLoader(train_
dataset,batch_size=args.batch_size,      shuffle=True,      num_
workers=args.workers, pin_memory=True, drop_last=True)
        eval_loader  =  torch.utils.data.DataLoader(val_dataset,
batch_size=1,   shuffle=False,   num_workers=args.workers,   pin_
memory=False)
        test_loader  =  torch.utils.data.DataLoader(test_dataset,
batch_size=1,   shuffle=False,   num_workers=args.workers,   pin_
memory=False)
    ##### train #####
    train_seg(args,train_result_path,train_loader,eval_loader)
    ##### test #####
            model_best_path  =  osp.join(osp.join(train_result_
path,'model'),'model_best.pth.tar')
    args.model_path = model_best_path
    test_seg(args,test_result_path,test_loader)
if __name__ == '__main__':
    main()
```

## References

Bochkovskiy, A., Wang, C.-Y., & Liao, H.-Y. M. (2020). Yolov4: Optimal speed and accuracy of object detection. arXiv preprint, arXiv:2004.10934

Canny, J. (1986). A computational approach to edge detection. *IEEE Transactions on Pattern Analysis and Machine Intelligence* (6), 679–698.

Chen, F. (2008). Image segmentation technology based on region growing method. *Technology Information* (15), 58–59.

Chen, L.-C., Zhu, Y., Papandreou, G., Schroff, F., & Adam, H. (2018). Encoder-decoder with atrous separable convolution for semantic image segmentation. In *Proceedings of the European Conference on Computer Vision (ECCV)* (pp. 801–818).

Chen, Y., Rohrbach, M., Yan, Z., Shuicheng, Y., Feng, J., & Kalantidis, Y. (2019). Graph-based global reasoning networks. In *Proceedings of the IEEE/CVF Conference on Computer Vision and Pattern Recognition* (pp. 433–442).

Choi, W., & Cha, Y.-J. (2019). SDDNet: Real-time crack segmentation. *IEEE Transactions on Industrial Electronics, 67*(9), 8016–8025.

DeVries, T., & Taylor, G. W. (2017). Improved regularization of convolutional neural networks with cutout. arXiv preprint, arXiv:1708.04552

Gao, J., Ren, M., & Tang, Z. (2003). Automatic road crack detection and identification. *Computer Engineering, 3*(2), 149–150.

Gu, Z., Cheng, J., Fu, H., Zhou, K., Hao, H., Zhao, Y., & Liu, J. (2019). Ce-net: Context encoder network for 2d medical image segmentation. *IEEE Transactions on Medical Imaging, 38*(10), 2281–2292.

He, F., Luo, H., Ping, A., & Yao, X. (2020). Bridge crack extraction algorithm based on Hessian matrix enhancement and local crack connection. *Journal of Guizhou University, 37*(3), 69–74.

Ioffe, S., & Szegedy, C. (2015). Batch normalization: Accelerating deep network training by reducing internal covariate shift. In *International Conference on Machine Learning* (pp. 448–456).

Jiang, Z., Wang, X., & Zhang, J. (1997). Genesis and developing conditions of earth fissures in shanxi down-faulted basin belt. *Journal of China University of Mining & Technology, 26*(3), 74–78.

Kaddah, W., Elbouz, M., Ouerhani, Y., Alfalou, A., & Desthieux, M. (2020). Automatic darkest filament detection (ADFD): A new algorithm for crack extraction on two-dimensional pavement images. *The Visual Computer, 36*(7), 1369–1384.

Kingma, D. P., & Ba, J. (2014). Adam: A method for stochastic optimization. arXiv preprint, arXiv: 1412.6980

Kipf, T. N., & Welling, M. (2016). Semi-supervised classification with graph convolutional networks. arXiv preprint, arXiv:1609.02907

Li, J., Jin, P., Zhu, J., Zou, H., Xu, X., Tang, M., & Su, Y. (2021). Multi-scale GCN-assisted two-stage network for joint segmentation of retinal layers and discs in peripapillary OCT images. *Biomedical Optics Express, 12*(4), 2204–2220.

Li, L., Chan, P., & Lytton, R. L. (1991). Detection of thin cracks on noisy pavement images. *Transportation Research Record,* 131–135.

Li, L., Ma, W., Li, L., & Lu, C. (2019). Research on bridge crack detection algorithm based on deep learning. *Acta Automatica Sinica, 45*(9), 1727–1742.

Li, X., Yang, Y., Zhao, Q., Shen, T., Lin, Z., & Liu, H. (2020). Spatial pyramid based graph reasoning for semantic segmentation. In *Proceedings of the IEEE/CVF Conference on Computer Vision and Pattern Recognition* (pp. 8950–8959).

Liu, F., Xu, G., Yang, Y., Niu, X., & Pan, Y. (2008). Novel approach to pavement cracking automatic detection based on segment extending. In *IEEE 2008 International Symposium on Knowledge Acquisition and Modeling* (pp. 610–614).

Long, J., Shelhamer, E., & Darrell, T. (2015). Fully convolutional networks for semantic segmentation. In *Proceedings of the IEEE Conference on Computer Vision and Pattern Recognition* (pp. 3431–3440).

Luo, W., Li, Y., Urtasun, R., & Zemel, R. (2016). Understanding the effective receptive field in deep convolutional neural networks. In *Advances in neural information processing systems*.

Ma, W. (2019). *Research on bridge crack detection algorithm based on deep learning* (Master thesis), Shanxi Normal University.

Meng, L. (2011). *Study on the genetic mechanism of ground fissures in Shanxi faulted basin* (Ph.D. thesis), Chang'an University.

Ronneberger, O., Fischer, P., & Brox, T. (2015). U-net: Convolutional networks for biomedical image segmentation. In *International Conference on Medical Image Computing and Computer-Assisted Intervention* (pp. 234–241). Springer.

Subirats, P., Dumoulin, J., Legeay, V., & Barba, D. (2006). Automation of pavement surface crack detection using the continuous wavelet transform. In *IEEE 2006 International Conference on Image Processing* (pp. 3037–3040).

Wang, S., Wu, X., Zhang, Y., & Chen, Q. (2018). Image crack detection with fully convolutional network based on deep learning. *Journal of Computer-Aided Design & Computer Graphics, 30*(5), 859–867.

Weng, P., Lu, Y., Qi, X., & Yang, S. (2019). Pavement crack segmentation technology based on improved fully convolutional networks. *Computer Engineering and Application, 55*(16), 235–239.

White, T. (2016). Sampling generative networks. arXiv preprint, arXiv:1609.04468

Xu, W., Tang, Z., & Lv, J. (2013). Pavement crack detection based on image saliency. *Journal of Image and Graphics, 18*(1), 69–77.

Yang, J., Hu, X., & Li, Y. (1999). Development of ground fissures and soil erosion in the Shanxi Graben System. *Research of Soil and Water Conservation, 6*(4), 10–14.

Yang, M. (2019). *Research on tunnel crack identification and analysis algorithm based on deep learning cascading technology* (Master thesis), Beijing University of Posts and Telecommunications.

Zhang, A., Wang, K. C., Li, B., Yang, E., Dai, X., Peng, Y., et al. (2017a). Automated pixel-level pavement crack detection on 3D asphalt surfaces using a deep-learning network. *Computer-Aided Civil and Infrastructure Engineering, 32*(10), 805–819.

Zhang, H., Cisse, M., Dauphin, Y. N., & Lopez-Paz, D. (2017b). Mixup: Beyond empirical risk minimization. arXiv preprint, arXiv:1710.09412

Zhang, L., Yang, F., Zhang, Y. D., & Zhu, Y. J. (2016). Road crack detection using deep convolutional neural network. In *2016 IEEE International Conference on Image Processing (ICIP)* (pp. 3708–3712).

Zhang, X. (2020). *Research on pavement crack segmentation algorithm based on expanded residual network* (Master thesis), Chang'an University.

Zhao, H., Shi, J., Qi, X., Wang, X., & Jia, J. (2017). Pyramid scene parsing network. In *Proceedings of the IEEE Conference on Computer Vision and Pattern Recognition* (pp. 2881–2890).

Zou, Q., Zhang, Z., Li, Q., Qi, X., Wang, Q., & Wang, S. (2018). Deepcrack: Learning hierarchical convolutional features for crack detection. *IEEE Transactions on Image Processing, 28*(3), 1498–1512.

The manufacturer's authorised representative in the EU is Springer Nature Customer Service Centre GmbH, Europaplatz 3, 69115 Heidelberg, Germany. If you have any concerns regarding our products, please contact ProductSafety@springernature.com

Printed and bound by CPI Group (UK) Ltd, Croydon, CR0 4YY

30/09/2026

02225314-0007